Victor Dmitrievich Lakhno

High-Temperature Superconductivity

Also of Interest

Physical Metallurgy.
Metals, Alloys, Phase Transformations
Vadim M. Schastlivtsev and Vitaly I. Zel'dovich, 2022
ISBN 9783110758016, e-ISBN 9783110758023

Sustainable Products.
Life Cycle Assessment, Risk Management, Supply Chains, Eco-Design
Michael Has, 2022
ISBN 9783110767292, e-ISBN 9783110767308

X-Ray Structure Analysis
Theo Siegrist, 2021
ISBN 9783110610703, e-ISBN 9783110610833

Intelligent Materials and Structures
Haim Abramovich, 2021
ISBN 9783110726695, e-ISBN 9783110726701

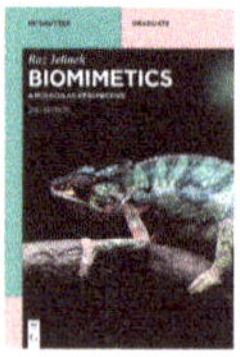

Biomimetics.
A Molecular Perspective
Raz Jelinek, 2021
ISBN 9783110709445, e-ISBN 9783110709490

Victor Dmitrievich Lakhno

High-Temperature Superconductivity

Bipolaron Mechanism

DE GRUYTER

Author
Prof. Dr. Victor Dmitrievich Lakhno
Institute of Mathematical Problems of Biology
Russian Academy of Sciences
Professor Vitkevich St. 1
142290 Pushchino, Moscow Region
Russia
lak@impb.ru

ISBN 978-3-11-078663-7
e-ISBN (PDF) 978-3-11-078666-8
e-ISBN (EPUB) 978-3-11-078668-2

Library of Congress Control Number: 2022936827

Bibliographic information published by the Deutsche Nationalbibliothek
The Deutsche Nationalbibliothek lists this publication in the Deutsche Nationalbibliografie; detailed bibliographic data are available on the Internet at http://dnb.dnb.de.

Cover image: Rost-9D/iStock/Getty Images Plus
Typesetting: Integra Software Services Pvt. Ltd.
Printing and binding: CPI books GmbH, Leck

www.degruyter.com

Preface

The monograph presents the polaron theory of superconductivity. The polaron theory, which was initiated by Landau's 1933 article, which appeared at the dawn of the formation of quantum mechanics, was first developed in the Soviet Union in the works by Pekar, and later, starting with the works by Fröhlich, in the publications of other theorists too. The reason for this interest was the fact that the polaron, being an elementary quasiparticle, provides the simplest meaningful example of a nonrelativistic quantum field theory. During almost 90-year development of polaron theory, practically the entire apparatus of theoretical physics has been tested. For example, the technique of continual integration became generally recognized after its successful application in the polaron theory by Feynman (1955). The method of collective coordinates in the quantum field theory also appeared after the work by Bogolyubov and Tyablikov on the strong coupling polaron. The methods of the theory of coherent and squeezed states, widely used in quantum optics and many other branches of physics, after the appearance of Glauber's works, were also first developed by Tulub in the theory of polarons. However, despite the efforts of many generations of theorists, the polaron problem has not yet found its complete solution.

The polaron theory, based on the Schrödinger equation for an electron interacting with an infinite number of field oscillators, is the simplest example of quantum field theory.

The mathematical apparatus for the formulation of such a quantum field problem, in principle, made it possible to do this immediately after the matrix formulation of Schrödinger's quantum mechanics by Heisenberg.

In the theory of solids, this was first done by Pekar for a strong coupling polaron. In modern terms, the solution to the problem was given in the form of an ansatz, in which it is presented in a multiplicative form, composed of the product of the electron wave function and the coherent states of the field oscillators. Coherent states are understood, according to their definition introduced later by Glauber, as the eigenstates of the annihilation operator.

Later, Fröhlich, Lee, Low, and Pines constructed an ansatz, in which a solution was given in the form of the product of the Heisenberg operators eliminating the electronic coordinates and therefore making the theory translation invariant and the shift operators acting on the vacuum wave function of the Hamiltonian, which corresponded to the case of weak interaction electron with field oscillators.

Finally, 10 years later, in 1961, Tulub gave a solution to the problem in the form of an ansatz, in which for the first time appeared the squeezed operator acting on the vacuum wave function. Tulub's ansatz gave a solution in the entire range of variation of the coupling constant and, in cases of strong and weak coupling, reproduced the results of Pekar, Fröhlich, Lee, Low, and Pines.

https://doi.org/10.1515/9783110786668-202

A completely different problem, seemingly unrelated to the polaron problem, arose in 1911 after the Dutch scientist Kamerlingh Onnes (1911) discovered superconductivity, long before the creation of quantum mechanics.

The construction of the quantum theory of superconductivity took place with great difficulty. The phenomenological foundations of the theory were laid in the works of Tisza and London, and its peak was the construction of a phenomenological theory by Ginzburg and Landau almost 40 years after the discovery of superconductivity.

The discovery of the isotope effect in the middle of the last century turned out to be a clue for the construction of a microscopic theory of superconductivity. It became clear that the lattice is involved in this phenomenon, that is, the electron–phonon interaction. A microscopic theory based on electron–phonon interaction was developed by Bardeen, Cooper, and Schrieffer (BCS) in 1957. The theory was constructed in the approximation of weak electron–phonon interaction and described well the situation in metals.

For some time, after the BCS triumph, the theory of superconductivity and the theory of polarons were developed as two unrelated theories. Despite their mathematical elegance, in practical terms, they were of little interest: the theory of polaron had no practical applications at all, and the theory of superconductivity was unable to indicate the way to increase the temperature of the superconducting transition. For this reason, the discovery of high-temperature superconductivity in 1986 by Bednorz and Müller had little to do with theoretical advances. A qualitative guiding idea for one of the authors of the discovery, K. Müller, was his assumption about the bipolaron mechanism of high-temperature superconductivity. The discovery of Müller and Bednorz inspired researchers all over the world to search for superconductors operating at room temperature (at the time of writing this monograph, the record temperature was 15 °C for a mixture of hydrogen sulfide and methane obtained at a pressure of 2.5 million atmospheres).

The bipolaron theory of superconductivity currently provides one of the possible solutions to this problem. A great contribution to its development was made by the works by Vinetskii, Anderson, Alexandrov, Mott, Ranninger, and others. Most of these works were devoted to small-radius polarons and bipolarons.

In the 2010–2020s of our century, the author developed a theory of superconductivity based on translation-invariant large-radius polarons, capable of explaining a large set of experimental facts on high-temperature superconductors. The theory is based on the equivalence of the Bogolyubov transformation, which he used to construct the theory of superconductivity, and the squeezed operator, involved in the solution of the polaron and bipolaron problem. The theory developed provides a solution to the spectral problem both in the case of a polaron and a bipolaron. The solution of the spectral problem for a polaron and a bipolaron can find application in various fields of physics, but the most urgent is its use in the theory of superconductivity. Although this theory has been published in a number of articles and reviews, its systematic presentation in the form of a monograph seems appropriate to the author. At the same time, much attention is paid to a consistent explanation of the experimental results based on the minimum number of assumptions made and the parameters of the theory.

The results presented in the monograph may be of interest for a wide range of areas of physics of condensed systems, as well as high-energy physics.

The book can be useful not only for specialists but also for senior students and graduate students who specialize in condensed matter physics.

Victor Dmitrievich Lakhno, March 2022

Contents

Appendices

1 Introduction

Pluralitas non est sine necessitate
(entities should not be multiplied without necessity)
Occam

1.1 Background of the problem

The theory of superconductivity (SC) for ordinary metals is one of the finest and long-established branches of condensed matter physics which involves macroscopic and microscopic theories and derivation of macroscopic equations from a microscopic description (Lifshitz and Pitaevskii, 1980). In this regard, the theory at its core was presented in its finished form and its further development should imply detalization and consideration of special cases.

The situation changed after the discovery of high-temperature SC (HTSC) (Bednorz and Müller, 1986). Surprisingly, the correlation length in oxide ceramics turned out to be several orders of magnitude less than in ordinary metal superconductors and the width of the gap much larger than the superconducting transition temperature (Ginzburg, 2000). The current state of the theory and experiment is given in books and reviews (Kakani and Kakani, 2009; Tohyama, 2012; Kruchinin et al., 2011; Sinha and Kakani, 2002; Benneman and Ketterson, 2008; Schrieffer, 1999; Cooper and Feldman, 2011; Plakida, 2010; Askerzade, 2012; Gunnarsson and Rösch, 2008; Moriya and Ueda, 2000; Manske, 2004; Kresin et al., 2021).

Presently, the main problem is to develop a microscopic theory capable of explaining experimental facts which cannot be explained by the standard Bardeen–Cooper–Schrieffer (BCS) theory (Bardeen et al., 1957).

While modern versions of a microscopic description of HTSC are many – phonon, plasmon, spin, exciton, and so on – the central point of a microscopic theory is the effect of electron coupling (Cooper effect). Such "bosonization" of electrons further lies in the core of the description of their superconducting condensate.

The phenomenon of pairing in a broad sense is the formation of bielectron states and in a narrow sense, if the description is based on a phonon mechanism – the formation of bipolaron states. For a long time, this concept has been in conflict with a great correlation length or the size of Cooper pairs in the BCS theory. The same reason hindered the treatment of SC as a boson condensate (see footnote at page 1177 in Bardeen et al. (1957)). In no small measure, this incomprehension was caused by a standard idea of bipolarons as very compact formations.

The first indication of the fallacy of this viewpoint was obtained in the work of Keldysh and Kozlov (1967) where an analogy between the BCS and Bose–Einstein condensation (BEC) was demonstrated while studying the properties of a high-density exciton gas. The results of Keldysh and Kozlov (1967) enabled one to develop the idea of

https://doi.org/10.1515/9783110786668-001

a crossover, that is, passing on from the BCS theory which corresponds to the limit of weak electron–phonon interaction (EPI) to the BEC theory which corresponds to the limit of strong EPI (Eagles, 1969; Nozi`eres and Schmitt-Rink, 1985; Loktev, 1996; Randeria, 1997; Uemura, 1997; Drechsler and Zwerger, 1992; Griffin et al., 1996). It was believed that additional evidence in favor of this way is Eliashberg strong coupling theory (Eliashberg, 1960). According to Marsiglio and Carbotte (1991), in the limit of infinitely strong EPI, this theory leads to the regime of local pairs, though greatly different from the BEC regime (Micnas et al., 1990).

However, attempts to develop a crossover theory between BCS and BEC faced insurmountable difficulties. For example, it was suggested to develop a theory, with the use of a T-matrix approach, where the T-matrix of the initial fermion system would transform into the T-matrix of the boson system as the EPI enhances (Zwerger, 2012; Bloch et al., 2008; Giorgini et al., 2008; Chen et al., 2005; Ketterle and Zwierlein, 2007; Pieri and Strinati, 2000). However, this approach turned out to fail even in the case of heavily diluted systems. Actually, the point is that even in the limit when a system consists of only two fermions, one cannot construct a one-boson state out of them. In the EPI theory, this problem is known as the bipolaron one.

One reason why the crossover theory failed is as follows. Like the bipolaron theory, the BCS theory is based on the Fröhlich Hamiltonian. For this Hamiltonian, an important theorem of the analyticity of the polaron and bipolaron energy with respect to EPI constant is proved (Gerlach and Löwen, 1991). However, in the BCS theory an important assumption is made – a real matrix element is replaced by a model quantity which is a matrix element truncated from the top and from the bottom of the phonon momenta. This procedure is, by no means, fair. As it is shown in Lakhno (2013), in the bipolaron theory this leads to side effects – existence of a local energy level separated by a gap from the quasicontinuous spectrum (Cooper effect). This solution is isolated and nonanalytical with respect to the coupling constant. In the BCS theory, just this solution forms the basis for the development of the SC theory.

As a result, the theory developed and its analytical continuation – Eliashberg theory – distort the reality and, in particular, make it impossible to construct a theory on the basis of the BEC. Replacement of a real matrix element by its model analog enables one to perform analytical calculations completely. In particular, a replacement of a real interaction by a local one in the BCS enabled one to derive the phenomenological Ginzburg–Landau model which is also a local model (Gor'kov, 1959). Actually, the power of this approach can hardly be overestimated since it enabled one to get a lot of statements consistent with the experiment.

Another more important reason why the crossover theory failed is that vacuum in the polaron (bipolaron) theory with spontaneously broken symmetry differs from the vacuum in the translation-invariant (TI) polaron (TI bipolaron) theory in the case of strong interaction which makes it impossible for the Eliashberg theory to pass on to the strong coupling TI bipolaron theory.

In this book, we present an SC theory based on the EPI. There the BCS corresponds to the limit of weak EPI, and the case of strong EPI corresponds to a TI bipolaron SC theory where the SC phase corresponds to a TI bipolaron Bose condensate (Chapter 4).

The relevance of a monograph on a bipolaron mechanism of SC is caused by the following facts: (1) Most publications on bipolaron SC are devoted to small-radius polarons (SRP) (Alexandrov and Krebs, 1992), while in the past time, after the theory of SC on the basis of SRP had been criticized (Chakraverty et al., 1998; de Mello and Ranninger, 1997, 1999; Firsov et al., 1999), interest has shifted to large-radius polarons; (2) most of papers published in the past decades were devoted to magnet-fluctuation mechanisms of SC while more recent experiments where record T_c (under high pressure) were obtained, were performed on hydrogen sulfides and lanthanum hydrides where magnetic interactions are lacking but there is a strong EPI; (3) crucial evidence in favor of a bipolaron mechanism is provided by recent experiments (Zhou et al., 2019), which demonstrate the existence of pairs at temperatures higher than T_c. (4) Important evidence for the bipolaron mechanism of SC is experiments (Božović et al., 2016), where the number of paired states in HTSC was demonstrated to be far less than the total number of current carriers.

In Chapter 2, we outline the main ideas of TI polarons and bipolarons in 3D polar crystals. The presentation here follows the review (Lakhno, 2015b). As in the BCS theory, the description of the TI bipolaron gas is based on the EPI and Fröhlich Hamiltonian. A general form of the solution to the polaron (bipolaron) problem in the form of the Tulub ansatz is presented.

As distinct from the BCS theory where the correlation length greatly exceeds the mean distance between the pairs, in this review we deal with the opposite case when the correlation length is far less than the distance between the pairs. The theory is generalized to the case of excitons (Lakhno, 2021a).

Chapter 3 is devoted to the one-dimensional case of the Holstein polaron and bipolaron, which is relevant in modern HTSC theories based on the Hubbard model and its modifications (Lakhno and Sultanov, 2011; Lakhno, 2014, 2016a).

In Chapter 4, we give a general solution to the spectral problem for a polaron and a bipolaron. The solution obtained is used to describe the thermodynamic properties of a three-dimensional Bose condensate of TI bipolarons. The critical transition temperature, energy, heat capacity, and transition heat of the TI bipolaron gas are discussed. The influence of an external magnetic field on the thermodynamic characteristics of a TI bipolaron gas is considered. The presentation here follows the review (Lakhno, 2020a). Based on the results obtained in this chapter, the theory of the pseudogap phase is constructed (Lakhno, 2021b).

In Chapter 5, a comparison with experiment is given for such characteristics as the maximum value of the magnetic field intensity at which the existence of TI bipolaron condensate is possible, the London penetration depth and its temperature dependence (Lakhno, 2019b). A theory of the isotope effect for the superconducting transition temperature and for the London penetration depth is developed (Lakhno, 2020b). The results obtained are used to explain experiments on HTSCs. Particular attention is paid to the fact that, according to the TI bipolaron theory of HTSC, different types of experiments measure different quantities as a SC gap. It is shown that tunneling experiments are used to determine the energy of bipolarons, while the angular resolution photoemission spectroscopy (ARPES) measures the phonon frequency for which the EPI is maximum. According to the TI bipolaron theory of superconductors, a natural explanation is given to such phenomena as the presence of kinks in spectral measurements of the gap, the angular dependence of the gap, the absence of the isotope effect for optimally doped superconductors, the presence of a pseudogap, and so on (Lakhno 2020a).

In Chapter 6, the TI bipolaron theory is used to describe a moving Bose condensate. The Little–Parks effect is considered.

Chapter 7 establishes a correlation between the theories of SC based on the concept of charge density waves (CDW) and TI bipolaron theory. It is shown that CDWs are formed in the pseudogap phase from bipolaron states due to Kohn anomaly, forming a pair density wave (PDW), with wave vectors corresponding to nesting. Formed in the pseudogap phase, CDWs coexist with SC at temperatures below the superconducting transition temperature, and their amplitude decreases with the formation of a Bose condensate from TI bipolarons, vanishing at a temperature equal to zero (Lakhno, 2020c, 2021c). The questions of CDW pinning and the theory of strange metals are considered.

1.2 Polaron and fundamental problems of nonrelativistic quantum field theory

The polaron theory is based on the Fröhlich Hamiltonian which describes the interaction of an electron with phonon field:

$$H = \frac{\hat{p}^2}{2m} + \sum_k \hbar\omega(k) a_k^+ a_k + \sum_k V_k \left[a_k e^{ikr} + a_k^+ e^{-ikr} \right], \tag{1.2.1}$$

where r is the radius vector of an electron, and $\hat{p}$ is its momentum; m is the electron effective mass; a_k^+, a_k are operators of the birth and annihilation of the field quanta with energy $\hbar\omega(k)$, and V_k is the matrix element of an interaction between an electron and a phonon field.

In the condensed matter physics, the polaron theory is a broad field which involves the description of electron properties of ionic crystals (Pekar, 1954; Fröhlich et al., 1950; Kuper and Whitfield, 1963; Devreese and Alexandrov, 2009; Lakhno, 1994), polar superconductors (Devreese, 1972; Devreese and Peeters, 1984), conducting polymers (Heeger et al., 1988; Ribeiro et al., 2013; Junior and Stafström, 2015), biopolymers (Schuster, 2004; Starikov et al., 2006), HTSCs (Emin, 1986; Emin, 2013; Alexandrov and Mott, 1996; Iadonisi et al., 2006), magnetic semiconductors (Nagaev, 1979; Lakhno, 1984), and other important objects of condensed matter.

The reason of such popularity of the polaron model is its universality. Fundamentally, all the physical phenomena are described relying on the quantum-field formulation. In nonrelativistic physics, its simplest realization is based on the use of Fröhlich Hamiltonian (1.2.1).

Various expressions for V_k and $\omega(\mathrm{k})$ in the case of ionic crystals, piezoelectrics, superconductors, nuclear matter, and degenerated semiconductor plasma are given in Lakhno and Chuev (1995). Thus, Hamiltonian (1.2.1) describes the motion of an electron in an ionic crystal if:

$$V_k = \frac{e}{|k|}\sqrt{\frac{2\pi\hbar\omega_k}{V\tilde{\varepsilon}}}, \qquad \omega_k = \omega_0, \tag{1.2.2}$$

where e is an electron charge and $\tilde{\varepsilon}$ is the effective dielectric constant.

In a piezoelectric semiconductor:

$$V_k = \frac{1}{k^{1/2}}\left(\frac{4\pi}{V}\frac{\hbar/\mu s}{\hbar^2\varepsilon/\mu e^2}\frac{\left\langle e_{ijk}^2\right\rangle}{2\varepsilon c}\right), \qquad \omega_k = sk, \tag{1.2.3}$$

where s is the sound velocity, c is the elastic constant, and $\left\langle e_{ijk}^2\right\rangle$ is the averaged square of the piezoelectric tensor.

In a nonpolar medium:

$$V_k = Gk\sqrt{\frac{\hbar}{2\rho V\omega_k}}, \qquad \omega_k = sk, \tag{1.2.4}$$

where ρ is the medium density and G is a certain constant. A similar (1.2.4) Hamiltonian is used in the SC theory.

In the case of nuclear matter:

$$V_k = \frac{g}{\sqrt{2\omega_k V}}, \qquad \omega_k = \sqrt{\frac{\mu_0^2 c^4}{\hbar^2} + c^2k^2}, \tag{1.2.5}$$

where g is a coupling constant of a nucleon with a meson field, μ_0 is a meson mass, and c is the light velocity.

In the case of interaction with the plasma of a degenerate semiconductor:

$$V_k = \sqrt{\frac{2\pi\omega_p\varepsilon(k)}{V\varepsilon_0(\varepsilon(k)-1)}}\frac{e}{|k|}, \qquad \omega_k = \omega_p\sqrt{\frac{\varepsilon(k)}{(\varepsilon(k)-1)}}, \tag{1.2.6}$$

where ω_p is a plasma frequency and $\varepsilon(k)$ is a plasma permittivity.

Recently, Hamiltonian (1.2.1) has been used to describe impurity atoms placed in a Bose–Einstein condensate of ultracold atoms (Grusdt et al., 2017), electrons in low-dimensional systems (Iadonisi et al., 2006; Jackson and Platzman, 1981; Chatterjee and Mukhopadhyay, 2018; Shikin and Monarkha, 1973), and so on.

For an impurity in Bose condensate, according to Cucchietti and Timmermans (2006), Sacha and Timmermans (2006), and Tempere et al. (2009):

$$V_k = \frac{a_{IB}\sqrt{n_0}}{\sqrt{2\pi}M}\left[\frac{(\xi k)^2}{2+(\xi k)^2}\right]^{1/4}, \quad \omega_k = ck\left[1+\frac{(\xi k)^2}{2}\right]^{1/2}, \tag{1.2.7}$$

c and ξ are the sound velocity in the condensate and a characteristic length of damping of disturbances in the condensate, respectively, a_{IB} is the length of boson scattering on an impurity, n_0 is a density of Bose condensate, $M^{-1} = m_B^{-1} + m_I^{-1}$, m_B. is a mass of a lattice atom, m_I is a mass of an impurity atom.

Rather, a simple form of Hamiltonian (1.2.1) has encouraged researchers to find an exact solution of the polaron problem. In the stationary state, an exact solution would give a spectrum of Hamiltonian (1.2.1) and, as a consequence, a solution of a wide range of condensed matter physics problems. However, the problem turned out to be much more complicated than it seemed to be. To solve it, various methods and techniques of the quantum field theory were used such as the Green function method, diagram technique, path integral method, renormalization group method, quantum Monte Carlo method, diagram Monte Carlo method, and so on. Various variational approaches, the most efficient of which turned out to be Feynman's path integral method, enabled researchers to find an approximate dependence of the polaron ground-state energy over the whole range of variation of the EPI constant α.

The above approaches, however, failed to determine the spectrum of Hamiltonian (1.2.1) even in the weak coupling limit (Tkach et al., 2015).

In the limit of strong coupling, in order to investigate the properties of Hamiltonian (1.2.1), starting with pioneering works by Pekar (1954), use was made of the canonical transformation:

$$a_k \to a_k - V_k\rho_k^* / \hbar\omega(k), \tag{1.2.8}$$

where $\rho_k{}^*$ is the Fourier component of the charge distribution density. Transformation (1.2.8) singles out the classical component (second term in the right-hand side of (1.2.8)), from the quantum field which, by assumption, should make the main contribution into the strong coupling limit. Starting with the work by Lieb and Yamazaki (1958) (see also Lieb and Thomas, 1997 and references therein), a lot of papers dealt with the proof that the functional of Pekar total energy for the polaron ground state yielded by (1.2.8) is asymptotically exact in the strong coupling limit.

In other words, it was argued that the choice of a variation wave function of the ground state in the form:

$$|\Psi\rangle = \phi(r)\exp\sum_k V_k \frac{\rho_k^*}{\hbar\omega(k)}(a_k - a_k^+)\ |0\rangle, \tag{1.2.9}$$

where $\phi(r)$ is the electron wave function, which, in the case of the total energy of the polaron ground state $E = \langle\Psi|H|\Psi\rangle$, leads to Pekar functional of the strong coupling:

$$E = \frac{\hbar^2}{2m}\int |\nabla\Psi|^2 d^3r - \sum_k \frac{V_k^2}{\hbar\omega(k)}\rho_k^*\rho_k, \tag{1.2.10}$$

yields rather an exact solution in the strong coupling limit.

In this case, the spectrum of the polaron excited states was considered only for a resting polaron $P = 0$, where P is the polaron momentum. Variation of (1.2.4) over Ψ^* leads to a nonlinear Schrödinger equation for the wave function Ψ, which has the form of Hartree equation. Numerical integration of this equation was performed in Balabaev and Lakhno (1980) and some polaron excited states and relevant renormalized phonon modes were found (see, for example, review by Lakhno and Chuev (1995)).

Hence, most of the papers on the polaron theory in the strong coupling limit realize the method of quantizing in the vicinity of a classical solution which is now widely used in the nonperturbative quantum field theory (Rajaraman, 1982).

Fundamentally, this method seemed unsatisfactory even at the early stages of the development of the polaron theory. Indeed, if in the strong coupling limit the polarization field can be considered to be classical and nonzero, it becomes unclear how this macroscopic state can be held by a single electron. It is possible only in the case when the field is equal to zero except for a small region where the electron is localized forming a self-consistent state with the field. In this case, the initial translational symmetry turns out to be broken: the polarization potential well can spontaneously form, equally likely, in any region of space. All the attempts to construct a TI theory on the basis of this physical picture yielded the same results that the initial semiclassical strong coupling theory developed by Pekar (1954).

The situation changed radically after the publication of the papers (Lakhno, 2010b, 2012a, 2012b, 2013; Kashirina et al., 2012) where a fundamentally different mechanism of an electron motion in a polar crystal was considered. According to Lakhno (2010b, 2012a, 2013) and Kashirina et al. (2012), when moving along a crystal, an electron not only displaces equilibrium states of atoms, but also alters the profile of their potential energy in the crystal which is equivalent to the formation of their squeezed oscillatory states (Hakioglu et al., 1995; Shumovskii 1991; Braunstein, 2005). For a TI polaron, average (i.e., classical) displacements of atoms from their equilibrium positions, as distinct from a Pekar polaron, are equal to zero.

Accordingly, polarization of the crystal is equal to zero too, since a TI polaron is spatially delocalized. However, the mean values of phonon occupation numbers in a polaron crystal are not equal to zero. This paradox is resolved by the fact that the nonzero mean number of phonons is caused by the availability of squeezed (i.e., nonclassical) states excited by an electron.

Squeezing of phonon states induced by the electron motion along a crystal leads to a new type of a bound state of the electron–phonon system described by a unified wave function which presents a new type of ansatz and cannot be presented as a factorized ansatz formed by the electron and phonon parts individually (Lakhno, 2015b).

The theory of squeezed states was first used in the polaron theory in Tulub (1961). In view of a nonoptimal choice of the variation wave function in Tulub (1961), the results obtained for the ground-state energy actually reproduce those derived by Pekar. This significantly delayed their use in the polaron theory. For this reason, intensive development of the squeezed state theory took place much later – after the paper by Glauber (1963), who drew attention to their important role in the understanding of the principle of uncertainty and the principle of superposition in quantum mechanics.

Presently, squeezed states have widespread application: in optics they are used to suppress self-noise of light; in computing technics, for the development of optical computers and communication lines; in precision measurements; in interference antennae of gravitation waves, and so on (see, for example, books and reviews: Teich and Saleh (1991), Schleich (2001),Misochko (2013)).

In the polaron theory, the squeezed state method, after the pioneering work by Tulub (1961), was used in Hang Zheng (1988a, 1988b, 1988c, 1988d, 1989) for a discrete model of a Holstein polaron (Holstein 1959a, 1959b) and in Porsch and Röseler (1967), Röseler (1968), Barentzen (1975), Kandemir and Altanhan (1994), Kandemir and Cetin (2005), Nagy (1991), Zhang Yan-Min and Ze Cheng (2007), Kervan et al. (2003) for a Fröhlich polaron. In Hang Zheng (1988a, 1988b, 1988c, 1988d, 1989), some very important results were obtained: first, the polaron ground-state energy of squeezed states turned out to be lower than that in all the papers on a Holstein polaron where an ordinary vacuum is used; second, the effective mass of a Holstein polaron calculated for squeezed states appeared to be much less than that obtained by Holstein (1959a).

This is not the case with Porsch and Röseler (1967), Röseler (1968), Barentzen (1975), Kandemir and Altanhan (1994), Kandemir and Cetin (2005), Nagy (1991), Zhang Yan-Min and Ze Cheng (2007), and Kervan et al. (2003), where the squeezed state theory was applied to Fröhlich Hamiltonian. Despite the fact that a considerable enhancing of polaron effects was observed when squeezed states were used, in general, the results did not differ from those obtained by Pekar (1954). As mentioned above, breakthrough results were obtained in Lakhno (2010b, 2012a, 2013) and Kashirina et al. (2012) where for Pekar–Fröhlich Hamiltonian, it was shown that the energy of a polaron ground state and the energy of a bipolaron for squeezed states is lower than that in the Pekar theory.

The most important application of the polaron and bipolaron theory is SC. Apparently, development of an SC theory is the most difficult problem of the condensed matter physics since it requires a solution of a multiparticle problem. This problem was solved by Bardeen, Cooper, and Schrieffer in the limit of weak interaction on the basis of Fröhlich Hamiltonian (1.2.1) (Bardeen et al., 1957). Its solution enabled one to explain some properties of ordinary superconductors.

The discovery of HTSC showed that the BCS theory probably cannot be applied to them since the EPI in HTSC materials cannot be considered to be weak. Presently, to describe this case, researchers resort to the use of Eliashberg theory (Eliashberg, 1960; Marsiglio and Carbotte, 1991; Carbotte, 1990), since it was developed for the case of a strong EPI and in the weak coupling limit it coincides with the BCS theory. In the absence of Coulomb repulsion, Eliashberg theory leads to the expression for T_c such that (Kresin and Wolf, 2009):

$$T_c = 0.25\omega_0 / \left(e^{2/\lambda} - 1\right)^{1/2}, \quad \lambda = 2\int \alpha^2(\omega) F(\omega)\omega^{-1} d\omega, \tag{1.2.11}$$

where $\alpha^2(\omega)$ determines the strength of EPI and depends on ω only slightly, $F(\omega)$ has the meaning of the density of phonon states. In the limit of weak coupling when $\lambda \to 0$ it follows from (1.2.11) that:

$$T_c = 0.25\omega_0 e^{-1/\lambda}, \tag{1.2.12}$$

that is the expression for T_c in BCS, while in the case of strong coupling when $\lambda \to \infty$:

$$T_c = 0.18\lambda^{1/2}\omega_0. \tag{1.2.13}$$

An important consequence of Eliashberg theory of strong coupling is the conclusion that there is no upper limit for T_c in the case of strong EPI.

However, the use of Eliashberg theory in the case of HTSC had limited success. This fact gave rise to a number of theories which were based not on the Hamiltonian of EPI (1.2.1) but on other types of interaction different from EPI. These works, eventually, faced the same problems that Eliashberg theory did.

In Eliashberg theory, a small parameter is a ratio ω/E_F, where ω is the phonon frequency, E_F is the Fermi energy. If $\omega/E_F \to 0$, then the EPI constant $\alpha \sim \omega^{-1/2} \to \infty$. The perturbation theory with respect to this parameter is developed for ordinary vacuum phonon functions $|0\rangle$, which are taken as a zero approximation. But for $\alpha \to \infty$ the proper choice of the zero approximation will be the function $\Lambda_0|0\rangle$,

$$\Lambda_0 = C\left\{\exp\frac{1}{2}\sum_{k,k'} a_k^+ A_{kk'} a_{k'}^+\right\}, \tag{1.2.14}$$

where Λ_0 is the operator of a squeezed state (Tulub 1961, 1960, 2015; Lakhno, 2015b). Hence, the Eliashberg theory developed for ordinary vacuum will give different results than the theory developed for squeezed vacuum does. Obviously, in the limit of weak coupling, when $\alpha \to 0$ the results of both the theories will coincide. However, as α increases, boson vacuum determined by the vacuum function $|0\rangle$ will be more and more unstable and for a certain critical value α_c, a new boson vacuum determined by the function $\Lambda_0|0\rangle$ emerges. It will be a lower energy state. The inapplicability of the Migdal theorem was probably first mentioned by Alexandrov and Krebs (1992), who, in relation to the SC theory based on SRP, pointed out that vacuum chosen on the basis of Migdal theorem "knows nothing" about another vacuum which is a polaron narrowing of the conductivity band and formation of a SRP in a new vacuum of squeezed states (Hang Zheng 1988a, 1988b, 1988c, 1988d, 1989). For this reason, Eliashberg theory is inadequate for the explanation of HTSC.

The foundation of superfluidity was laid in papers by London (1938) and Tisza (1938), who were the first to relate the fundamental phenomenon of BEC to phenomenon of superfluidity. The idea to treat SC as superfluidity of electron liquid was suggested by Landau (Landau, 1941; Landau 1947). There a spectrum of elementary excitations of a superfluid liquid was introduced for the first time which received the name of a roton spectrum and enabled one to construct a statistical mechanics of a superfluid state. Landau could not transfer the ideas of his work on superfluidity to SC because of a difference in statistics: Fermi statistics for electrons in metals and Bose statistics for helium atoms in liquid helium II. The work by Bogolyubov (1947) on superfluidity which related the phenomenon of Bose condensate to superfluidity could have accelerated the construction of the SC theory on the basis of Bose condensate, however, at that time there was not an example of Bose gas of charged bosons which is necessary for SC.

Further sequence of events is well known: Ginzburg and Landau (1950) developed a phenomenological theory of SC in which a microscopic mechanism of SC was not discussed since its possible nature was unclear.

Finally, in the work by Bardeen et al. (1957), a microscopic mechanism of SC was found. This was the mechanism of Cooper pairing of electrons. Cooper pairs, being bosons, supposedly could have played the role of particles from which Bose gas consists and, thus, have combined the theories of SC and superfluidity. However, that did not happen. The answer was given in the BCS theory per se – the size of Cooper pairs in metals turned out to be so huge that in each pair there was about of a million of other pairs. For this reason, an analogy between a Bose–Einstein condensate and SC, was discarded in Bardeen et al. (1957). Interest in it emerged only in 1986 when Müller and Bednorz discovered HTSC.

To be fair, it should be emphasized that a possibility of the formation of such a singular quantum state as Bose condensate was predicted by Einstein on the basis of generalization of Bose statistics to the case of a finite mass of a Bose particle. Until the publication of the BCS work, there was not an example of a charged boson with a finite mass in the condensed matter physics. The first example of a possible existence of such quasiparticles was a Cooper pair which enabled BCS to construct a theory of SC. A Cooper pair, as was mentioned above, being overlapped with others could not be a true quasiparticle. For the same reason, both in the BCS and in Bogolyubov (1958) theory, there is only a single-electron spectrum of Fermi-type excitations. Hence in the framework of BCS, as it was stated by its authors, a theory of Bose condensate cannot be constructed.

In 1970–1980, a small-radius bipolaron (SRBP) was considered as a quasiparticle possessing the properties of a charged boson, having a mass, and capable of forming a Bose condensate in narrow-bandgap crystals (Anderson, 1975).

For a long time, works on SC based on the idea of Bose condensate of SRBP were developed by Alexandrov and Krebs (1992), Alexandrov and Mott (1996), Alexandrov and Kornilovitch (1999), Alexandrov and Ranninger (1981), and Firsov et al. (1999).

In view of a large mass of SRP and SRBP, the temperature of the SC transition determined by the temperature of BEC formation should be low. This fact was pointed out in Chakraverty et al. (1998) and de Mello and Ranninger (1997, 1999), which criticized the SC theory based on SRP.

After the discovery of HTSC some other approaches were developed the most popular of which was Anderson resonating valence bond theory and t-J model (Anderson, 1997; Izyumov, 1997).

Notwithstanding a strong attraction of these models from the viewpoint of theory, for example, a possibility to describe both conducting and magnetic properties of crystals on the basis of one simple Hamiltonian, they turned out to be ineffective for explaining HTSC. In particular the fact of a possible existence of a SC phase in these models did not receive a reliable proof.

In view of the fact that recent experiments by Zhou et al. (2019) and Božović et al. (2016) suggest a phonon nature of the SC mechanism in HTSC with a record T_c, further presentation is based on EPI. Being general, the theoretical approaches considered can be applied to other types of interaction different from EPI.

2 Pekar's ansatz and the strong coupling problem in polaron theory

A detailed consideration is given to the translation-invariant theory of Tulub polaron constructed without the use of Pekar's ansatz. It is shown that the wave function of a polaron (bipolaron) is the product of the unitary Heisenberg operators, the shift operator and the squeezed operator acting on the vacuum wave function. A fundamental result of the theory is that the value of the TI polaron energy is lower than that obtained on the basis of Pekar's ansatz which was considered as an asymptotically exact solution in the strong coupling limit. In the case of bipolarons the theory yields the best values of the coupling energy and critical parameters of their stability. Numerous physical consequences of the existence of translation-invariant polarons and bipolarons are discussed. The theory is generalized to the case of excitons.

2.1 Introduction: Pekar's ansatz

As is known, polaron theory was among the first to describe the interaction between a particle and a quantum field. Various aspects of the polaron theory are presented in numerous reviews and books (Pekar, 1954; Kuper and Whitfield, 1963; Devreese, 1972; Devreese and Peeters, 1984; Devreese and Alexandrov, 2009; Firsov, 1975; Lakhno, 1994; Kashirina and Lakhno, 2010, 2013; Emin, 2013). Being nonrelativistic, the theory does not contain any divergencies and for more than 60 years has been a testing area for approbation of various methods of the quantum field theory. Though no exact solution of the polaron problem has been found up to now, it has been believed that the properties of the ground state are known in detail. This primarily refers to the limit cases of weak and strong coupling. A solution in the weak coupling limit was given by Fröhlich et al. (1950), and that in the strong coupling one was found by Pekar (1954, 1946a, 1946b). By now rather an exact solution has been obtained for the energy of the polaron ground state in the weak coupling limit (Smondyrev, 1986; Selyugin and Smondyrev, 1989):

$$E_0 = -(\alpha + 0.0159196220\alpha^2 + 0.000806070048\alpha^3 + \cdots)\hbar\omega_0 \tag{2.1.1}$$

where $\hbar\omega_o$ is the energy of an optical phonon and α is a constant of electron–phonon coupling.

A solution of the problem in the opposite strong coupling limit was given by Pekar on the assumption that the wave function Ψ of the electron + field system has the form:

$$\Psi(r, q_1, \ldots, q_i, \ldots) = \psi(r)\Phi(q_1, \ldots, q_i, \ldots), \tag{2.1.2}$$

where $\psi(r)$ is the electrons wave function depending only on the electron coordinates and Φ is the wave function of the field depending only on the field coordinates.

https://doi.org/10.1515/9783110786668-002

Pekar (1954) considered ansatz (2.1.2) to be an approximate solution. In the pioneer works by Bogolyubov (1950) and Tyablikov (1951), it was shown that in a consistent translation-invariant theory the use of ansatz (2.1.2) (for decomposed coordinates introduced in Bogolyubov (1950)and Tyablikov (1951)) gives the same results for the polaron ground state energy as the semiclassical Pekar (1954, 1946) theory does. Since the problem of the translational invariance of the polaron was first clearly indicated in these works, let us dwell in more detail on the assumptions made in them.

For this purpose, we pass in the Fröhlich Hamiltonian (1.2.1) from the creation and annihilation operators a_k^+, a_k to the complex coordinates of the field q_k using the relations:

$$\begin{aligned} q_k &= \left(a_k + a^+_{-k}\right)/\sqrt{2} \\ -\frac{\partial}{\partial q_k} &= \left(a_k^+ - a_{-k}\right)/\sqrt{2}. \end{aligned} \tag{2.1.3}$$

As a result, Hamiltonian (1.2.1) takes the form of the Pekar Hamiltonian:

$$\begin{aligned} &H = \frac{\hat{p}^2}{2m} + \sum_k A_k q_k e^{ikz} + \frac{1}{2}\sum \hbar\omega(k)\left(q_k q_{-k} - \frac{\partial}{\partial q_k}\frac{\partial}{\partial q_{-k}}\right), \\ &A_k = \sqrt{2}C_k \end{aligned} \tag{2.1.4}$$

It can be seen from (2.1.4) that the coordinates q_k and momenta $-i\partial/\partial q_k$ canonically conjugate to them enter the field energy in the same way. To make their contribution into energy unequal, in the works by Bogolyubov (1950) and Tyablikov (1951), the transformation is used:

$$\tilde{q}_k = \varepsilon q_k, \qquad \omega(k) = \varepsilon^2 \nu(k) \tag{2.1.5}$$

As a result, Hamiltonian (2.1.4) takes on the form:

$$\begin{aligned} &H = \frac{\hat{p}^2}{2m} + \sum_k \tilde{A}_k \tilde{q}_k + \frac{1}{2}\sum_k \nu(k)\tilde{q}_k\tilde{q}_{-k} - \frac{\varepsilon^4}{2}\sum_k \nu(k)\frac{\partial}{\partial \tilde{q}_k}\frac{\partial}{\partial \tilde{q}_{-k}}, \\ &\tilde{A}_k = \frac{e}{k}\sqrt{\frac{2\pi\hbar c\nu(k)}{V}} = \frac{1}{\varepsilon}A_k, \end{aligned}$$

in which the small parameter ε is involved only in the last term of the Hamiltonian, whose contribution in the zero approximation can be neglected.

It is easy to see, however, that through the transformation $q_k = \varepsilon\tilde{q}_k$, $\omega = \varepsilon^2\nu$ we get that the potential energy of the field is small, which is now $\sim\varepsilon^4$ and the kinetic energy should be considered as a zero approximation.

In fact, the reason for the choice in the works by Bogolyubov (1950) and Tyablikov (1951) of the potential energy as the zero approximation is associated with additional assumptions. To reproduce the Pekar's limit of the strong coupling in Bogolyubov

(1950) and Tyablikov (1951), an assumption is made about the existence of the classical component of the quantum field. The presence of such a component does not in any way affect the kinetic energy of the field, while its contribution to the potential energy of the field becomes dominant. Up to date the energy of ground state with the use of (1.2) was found with great accuracy and is equal (Miyake, 1975, 1976):

$$E = (-0.108513\alpha^2 - 2.836)\hbar\omega_0. \tag{2.1.6}$$

In fact, introducing the classical component is equivalent to choosing a solution in the form of the Pekar's ansatz (2.1.2). When the assumption of the presence of the classical component of the field in the theory by Bogolyubov (1950) and Tyablikov (1951) is abandoned, the small parameter also ceases to exist. This, in particular, is associated with all unsuccessful attempts to construct a perturbation theory on ε in the strong-coupling limit without resorting to the concept of the classical field.

The concept of Pekar's ansatz (2.1.2) as an exact solution to the problem of a strong-coupling polaron was fully confirmed after the publication of Adamowski et al. (1980), in which the method of path integrals, that is, without the use of ansatz (2.1.2), asymptotics (2.1.3) was rigorously proved (see also the review by Gerlach and Löwen (1991)).

Before the publication of Adamowski et al. (1980), many attempts were made to improve the strong coupling theory (Höhler, 1955; Allcock, 1956; Gross, 1955; Buymistrov and Pekar, 1957; Toyozawa, 1961; Gross, 1976). The reason why Pekar's ansatz caused the feeling of disappointment was translation invariance of the initial polaron Hamiltonian. When ansatz for the wave function $\psi(r)$ in (2.1.2) is used, the wave equation has a localized solution. The electron is localized in a potential polarization well induced by it. In other words, the solution obtained does not possess the symmetry of the initial Hamiltonian. Self-trapping of the electron in the localized potential well leads to a spontaneous breaking of the systems symmetry. Attempts to restore the initial symmetry were based on the use of degeneration of the system with broken symmetry. Since in a homogeneous and isotropic medium nothing should depend on the position of the polaron well center r_0, one can "spread" the initially localized solution over all the positions of the polaron potential well by choosing the wave function in the form of a linear combination in all the positions of the well.

In the most consistent form, this program was carried out in Buymistrov and Pekar (1957). With this end in view for the wave function which is an eigenfunction of the total momentum, the authors used a superposition of plain waves corresponding to the total momentum multiplied by wave functions obtained from (2.1.2) to which a translations operator is applied. In other words, they took an appropriate superposition with respect to all the positions of the polaron well r_0.

The main result of Buymistrov and Pekar (1957) is that calculation of the polaron ground state energy with such a delocalized function yields the same value as calculations with localized function (2.1.2) do. Buymistrov and Pekar (1957) also reproduced the value of the polaron mass which was earlier obtained by Landau and Pekar (1948), on the assumption that polaron moves in medium in the localized state (2.1.2) (for acoustic 1D polaron see Schüttler and Holstein, 1986). The results derived in Buymistrov and Pekar (1957) were an important step in resolving the contradiction between the requirement that the translation-invariant wave function be delocalized while the wave function of the self-trapped state be localized.

Notwithstanding the success achieved with this approach it cannot be considered fully adequate since it has quite a few inconsistencies. They follow from the very nature of the semiclassical description used. Indeed, the superposition constructed in Buymistrov and Pekar (1957), on the one hand, determines the polaron delocalized state, but on the other hand, without changing this state, one can measure its position and find out a localized polaron well with an electron localized in it. The reason of this paradox is a classical character of the polaron well in the strong coupling limit and, as a consequence, commutation of the total momentum operator with the position of the polaron well[1] r_0. To remedy this defect some approaches were suggested in which the quantity r_0, which is not actually an additional degree of freedom was considered to be that with some additional constraints on the function $r_0(r, q_1, \ldots, q_i, \ldots)$. Discussion of these challenges associated with solution of the problem of introducing collective coordinates is given in Lakhno (1998).

Since the results obtained by introducing collective coordinates into the polaron theory are polemical it seems appropriate to describe strict results of the translation-invariant theory without recourse to the concept of collective coordinates. The aim of this review is to present an approach used in the strong coupling limit which does not use Pekar's ansatz.

A solution possessing these properties in the case of a strong coupling polaron was originally found by Tulub (1960, 1961). For nearly half a century, the result obtained in Tulub (1960, 1961) was not recognized by specialists working in the field of polaron theory. The reason why the importance of the result obtained in Tulub (1960, 1961) was not appreciated was an improper choice of the probe wave function in Tulub (1961) to estimate the ground state. As a result, the ground state energy was found in Tulub (1961) to be: $E_0 = -0.105\alpha^2\hbar\omega_0$, which is larger than in (2.1.6). An appropriate choice of the wave function has been made quite recently in

1 At the rise of quantum mechanics, the founders of the science were fully aware of the difficulties arising here. Thus, for example, Bethe (1964) notices that for a proper quantum-mechanical description of an interaction between a field and particles, quantizing of the field is required, that is, quantum theory of the field: "The fact is that, when quantizing mechanical parameters (coordinates and momenta) one should also quantize the associated fields. Otherwise, as Bohr and Rosenfeld (1933) showed, an imaginary experiment can be suggested which consists in simultaneous measurement of the coordinate and the momentum of a particle from examination of the field induced by it. This contravenes Heisenberg's Uncertainty Principle."

Kashirina et al. (2012). This has yielded a lower than in (2.1.6) value of the polaron ground state energy equal to: $E_0 = -0.125720\alpha^2\hbar\omega_0$. Hence, actually we have to do with inapplicability of adiabatic approximation in the case of a polaron, though it is fundamental for solid-state physics.

In this chapter we present the main points of the translation-invariant polaron (TI polaron) theory and generalize it to the case of a bipolaron and exciton.

2.2 Coordinate-free Hamiltonian: weak coupling

Let us proceed from Pekar–Fröhlich Hamiltonian:

$$H = -\frac{\hbar^2}{2m}\Delta_r + \sum_k V_k(a_k e^{ikr} + a_k^+ e^{-ikr}) + \sum_k \hbar\omega_k^0 a_k^+ a_k, \tag{2.2.1}$$

where a_k^+, a_k are operators of the birth and annihilation of the field quanta with energy $\hbar\omega_k^0 = \hbar\omega_0$, m is the electron effective mass, and V_k is a function of the wave vector k.

Electron coordinates can be excluded from (2.2.1) via Heisenberg (1930) transformation:

$$S_1 = \exp\left\{\frac{i}{\hbar}\left(\mathbf{P} - \sum_k \hbar k\mathbf{k} a_k^+ a_k\right)\mathbf{r}\right\}, \tag{2.2.2}$$

where $\mathbf{P}$ is the total momentum of the system. Application of S_1 to the field operators yields:

$$S_1^{-1} a_k S_1 = a_k e^{-ikr}, \quad S_1^{-1} a_k^+ S_1 = a_k^+ e^{ikr}$$

Accordingly, the transformed Hamiltonian $\tilde{H} = S_1^{-1} H S_1$ takes on the form:

$$\tilde{H} = \frac{1}{2m}\left(\mathbf{P} - \sum_k \hbar\mathbf{k} a_k^+ a_k\right)^2 + \sum_k V_k(a_k + a_k^+) + \sum_k \hbar\omega_k^0 a_k^+ a_k, \tag{2.2.3}$$

Since Hamiltonian (2.2.3) does not contain electron coordinates, it is obvious that solution of the polaron problem obtained on the basis of (2.2.3) is translation invariant. Lee et al. (1953) studied the ground state (2.2.3) with the probe wave function $|\Psi\rangle_{LLP}$:

$$|\Psi\rangle_{LLP} = S_2|0\rangle, \tag{2.2.4}$$

where

$$S_2 = \exp\left\{\sum_k f_k(a_k^+ - a_k)\right\}, \tag{2.2.5}$$

f_k are variational parameters having the meaning of the value of displacement of the field oscillators from their equilibrium positions and $|0\rangle$ is the vacuum wave function. The quantity f_k in S_2 (2.2.5) is determined by minimization of energy $E = \langle 0|S_2^{-1}\tilde{H}S_2|0\rangle$, which for $P = 0$ yields:

$$E = 2\sum_k f_k V_k + \frac{\hbar^2}{2m}\left[\sum_k \mathbf{k} f_k^2\right]^2 + \sum_k \frac{\hbar^2 k^2}{2m} f_k^2 + \sum_k \hbar\omega_k^0 f_k^2, \tag{2.2.6}$$

$$f_k = -\frac{V_k}{\hbar\omega_k^0 + \hbar^2 k^2/2m}. \tag{2.2.7}$$

In the case of an ionic crystal:

$$V_k = \frac{e}{k}\sqrt{\frac{2\pi\hbar\omega_0}{\tilde{\varepsilon}V}} = \frac{\hbar\omega_0}{ku^{1/2}}\left(\frac{4\pi\alpha}{V}\right)^{1/2}, \quad u = \left(\frac{2m\omega_0}{\hbar}\right)^{1/2}, \quad \alpha = \frac{1}{2}\frac{e^2 u}{\hbar\omega_0\tilde{\varepsilon}}, \quad \tilde{\varepsilon}^{-1} = \varepsilon_\infty^{-1} - \varepsilon_0^{-1}, \tag{2.2.8}$$

where e is an electron charge, ε_∞ and ε_0 are high-frequency and static dielectric permittivities, α is a constant of electron–phonon coupling. With substitution of (2.2.8) into (2.2.6) and (2.2.7) the ground state energy becomes $E = -\alpha\hbar\omega_0$, which is the energy of a weak coupling polaron in the first order with respect to α.

A solution of the problem of transition to the strong coupling case in coordinate-free Hamiltonian (2.2.3) was found on the basis of the general translation-invariant theory constructed in Tulub (1961). The main points of this theory are given in the next section.

2.3 Coordinate-free Hamiltonian: general case

To construct the general translation-invariant theory in Tulub (1960, 1961), use was made of a canonical transformation of Hamiltonian (2.2.3) with the use of operator S_2 (2.2.5), which leads to a shift of the field operators:

$$S_2^{-1} a_k S_2 = a_k + f_k, \quad S_2^{-1} a_k^+ S_2 = a_k^+ + f_k. \tag{2.3.1}$$

The resultant Hamiltonian $\tilde{\tilde{H}} = S_2^{-1}\tilde{H}S_2$ has the form:

$$\tilde{\tilde{H}} = H_0 + H_1, \tag{2.3.2}$$

where

$$H_0 = \frac{\mathbf{P}^2}{2m} + 2\sum_k V_k f_k + \sum_k \left(\hbar\omega_k^0 - \frac{\hbar\mathbf{k}\mathbf{P}}{m}\right) f_k^2 + \frac{1}{2m}\left(\sum_k \mathbf{k} f_k^2\right)^2 + H_{KB}, \tag{2.3.3}$$

$$H_{KB} = \sum_k \hbar\omega_k a_k^+ a_k + \frac{1}{2m}\sum_{k,k'} \mathbf{kk}' f_k f_{k'} \left(a_k a_{k'} + a_k^+ a_{k'}^+ + a_k^+ a_{k'} + a_{k'}^+ a_k\right), \tag{2.3.4}$$

$$\hbar\omega_k = \hbar\omega_k^0 - \frac{\hbar\mathbf{kP}}{m} + \frac{\hbar^2 k^2}{2m} + \frac{\hbar\mathbf{k}}{m}\sum_{k'} \hbar\mathbf{k}' f_{k'}^2. \tag{2.3.5}$$

Hamiltonian H_1 contains terms "linear," "triple," and "quadruple" in the birth and annihilation operators. With an appropriate choice of the wave function diagonalizing quadratic form (2.3.4), mathematical expectation H_1 becomes zero (Appendix A). In what follows we believe that $\hbar = 1$, $\omega_0 = 1$, $m = 1$. To transform H_{KB} to a diagonal form we put:

$$q_k = \frac{1}{\sqrt{2\omega_k}}\left(a_k + a_k^+\right), \quad p_k = -i\sqrt{\frac{\omega_k}{2}}\left(a_k - a_k^+\right), \quad \mathbf{z}_k = \mathbf{k} f_k \sqrt{2\omega_k}. \tag{2.3.6}$$

With the use of (2.3.6), expression (2.3.4) is written as

$$H_{KB} = \frac{1}{2}\sum_k \left(p_k^+ p_k + \omega_k^2 q_k^+ q_k\right) + \frac{1}{2}\left(\sum_k \mathbf{z}_k q_k\right)^2 - \frac{1}{2}\sum_k \omega_k. \tag{2.3.7}$$

This yields the following motion equation for operator q_k:

$$\ddot{q}_k + \omega_k^2 q_k = -\mathbf{z}_k \sum_k \mathbf{z}_{k'} q_{k'}. \tag{2.3.8}$$

Let us search for a solution of system (2.3.8) in the form:

$$q_k(t) = \sum_{k'} \Omega_{kk'} \xi_{k'}(t), \quad \xi_k(t) = \xi_k^0 e^{i\nu_k t}. \tag{2.3.9}$$

Relation between matrix $\Omega_{kk'}$ and Green function is considered in Appendix B. As a result, we express matrix $\Omega_{kk'}$ as follows:

$$\left(\nu_{k'}^2 - \omega_k^2\right)\Omega_{kk'} = \mathbf{z}_k \sum_{k''} \mathbf{z}_{k''} \Omega_{k''k'}. \tag{2.3.10}$$

Let us consider determinant of this system which is derived by replacing the eigenvalues ν_k^2 in (2.3.10) with the quantity s which can differ from ν_k^2. The determinant of this system will be:

$$\det\left|\left(s - \omega_k^2\right)\delta_{kk'} - \mathbf{z}_k \mathbf{z}_{k'}\right| = \prod_k \left(s - \nu_k^2\right). \tag{2.3.11}$$

On the other hand, according to Wentzel (1942):

$$\det\left|\left(s - \omega_k^2\right)\delta_{kk'} - \mathbf{z}_k \mathbf{z}_{k'}\right| = \prod_k \left(s - \omega_k^2\right)\left(1 - \frac{1}{3}\sum_{k'} \frac{\mathbf{z}_{k'}^2}{s - \omega_{k'}^2}\right)^3. \tag{2.3.12}$$

It is convenient to introduce the quantity $\Delta(s)$:

$$\Delta(s) = \prod_k \left(s - \nu_k^2\right) / \prod_k \left(s - \omega_k^2\right). \tag{2.3.13}$$

With the use of (2.3.11) and (2.3.12) $\Delta(s)$ is expressed as

$$\Delta(s) = \left(1 - \frac{1}{3}\sum_{k'} \frac{\mathbf{z}_{k'}^2}{s - \omega_{k'}^2}\right)^3. \tag{2.3.14}$$

From (2.3.11) and (2.3.12), it follows that the frequencies ν_k renormalized by interaction are determined by a solution to the equation:

$$\Delta\left(\nu_k^2\right) = 0. \tag{2.3.15}$$

The change in the systems energy ΔE caused by the electron-field interaction is equal to

$$\Delta E = \frac{1}{2}\sum_k (\nu_k - \omega_k). \tag{2.3.16}$$

To express the quantity ΔE via $\Delta(s)$ we use the Wentzel (1942) approach. Following Wentzel (1942), we write down the identity equation:

$$\begin{aligned} \sum_k \left\{f\left(\nu_k^2\right) - f\left(\omega_k^2\right)\right\} &= \frac{1}{2\pi i}\oint_C ds f(s) \sum_k \left(\frac{1}{s - \nu_k^2} - \frac{1}{s - \omega_k^2}\right) = \\ &= \frac{1}{2\pi i}\oint_C ds f(s) \frac{d}{ds}\ln\Delta(s) = -\frac{1}{2\pi i}\oint_C ds f'(s)\ln\Delta(s), \end{aligned} \tag{2.3.17}$$

where integration is carried out over the contour presented in Fig. 2.1.

Taking $f(s) = \sqrt{s}$, we get:

$$\Delta E = \frac{1}{2}\sum_k (\nu_k - \omega_k) = -\frac{1}{8\pi i}\oint_C \frac{ds}{\sqrt{s}}\ln\Delta(s). \tag{2.3.18}$$

Turning in (2.3.14) from summing up to integration with the use of the relation:

$$\sum_k = \frac{1}{(2\pi)^3}\int d^3k$$

in a continuous case, using for $\mathbf{z}_k$ expression (2.3.6) for $\Delta(s)$ we obtain

$$\Delta(s) = D^3(s), \quad D(s) = 1 - \frac{2}{3(2\pi)^3}\int \frac{k^2 f_k^2 \omega_k^2}{s - \omega_k^2} d^3k. \tag{2.3.19}$$

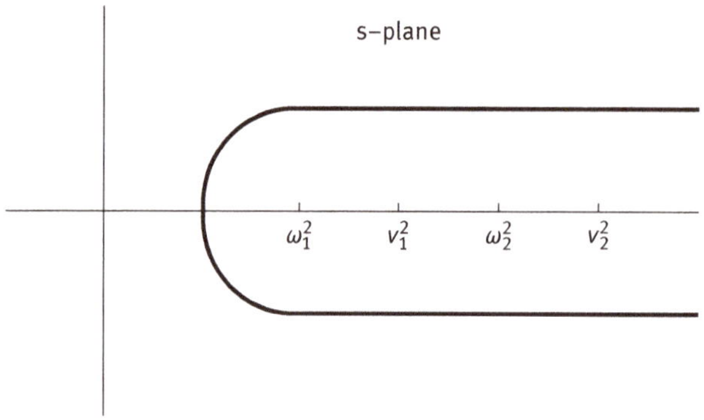

Fig. 2.1: Contour C.

As a result, the total energy of the electron is

$$E = \Delta E + 2\sum_k V_k f_k + \sum_k f_k^2 \omega_k^0. \tag{2.3.20}$$

The results obtained here are general and valid for various polaron models (i.e., any functions V_k and ω_k^0). In Sections 2.4 and 2.5, we consider limit cases of weak and strong coupling which follow from general expression (2.3.20). On the assumption that $\mathbf{P} \neq 0$ according to Tulub (1960), expression (2.3.20) takes the form:

$$E = \frac{P^2}{2m} + \Delta E(\mathbf{P}) + 2\sum_k V_k f_k + \sum_k f_k^2 \omega_k^0,$$

$$\Delta E(\mathbf{P}) = -\frac{1}{8\pi i}\oint_C \frac{ds}{\sqrt{s}} \ln \prod_{i=1}^{3} D^i(s),$$

$$D^i(s) = 1 - \sum_k \frac{(z_k^i)^2}{s - \omega_k^2},$$

where z_k^i is the ith component of vector $\mathbf{z}_k$. Notice that functions f_k, ω_k, and z_k should be considered to be dependent both on $|k|$ and on $\mathbf{kP}$.

2.4 Weak coupling limit in Tulub's theory

Quantities f_k in the expression for the total energy E (2.3.20) should be found from the minimum condition: $\delta E/\delta f_k = 0$, which yields the following integral equation for f_k:

$$f_k = -V_k/\left(1 + k^2/2\mu_k\right), \quad \mu_k^{-1} = \frac{\omega_k}{2\pi i}\oint_C \frac{ds}{\sqrt{s}} \frac{1}{(s - \omega_k^2)D(s)}. \tag{2.4.1}$$

In the case of weak coupling $\alpha \to 0$ and eq. (2.4.1) can be solved with the use of perturbation theory. In a first approximation, as $\alpha \to 0$ $D(s) = 1$ and μ_k^{-1} is equal to

$$\mu_k^{-1} = \frac{\omega_k}{2\pi i} \oint_C \frac{ds}{\sqrt{s}} \frac{1}{(s - \omega_k^2) D(s)} = 1. \tag{2.4.2}$$

Accordingly, f_k from (2.4.1) is written as follows:

$$f_k = -V_k / \left(1 + k^2/2\right). \tag{2.4.3}$$

The quantity ΔE involved in the total energy takes on the form:

$$\Delta E = -\frac{3}{8\pi i} \oint_C \frac{ds}{\sqrt{s}} \ln D(s), \quad \ln D(s) = -\frac{2}{3(2\pi)^3} \int \frac{k^2 f_k^2 \omega_k}{s - \omega_k^2} d^3k. \tag{2.4.4}$$

With the use of (2.4.3), integrals involved in (2.4.4) are found to be $\Delta E = (\alpha/2)\hbar\omega_0$. Having calculated the rest of the terms involved in expression (2.3.20), we get the first term of the expansion of polaron total energy in the coupling constant α: $E = -\alpha\hbar\omega_0$.

In Tulub (1961), Porsch and Röseler (1967), and Röseler (1968), a general scheme of calculating the higher terms of expansion in α was developed. In particular, the eigen energy and effective mass were found to be (Röseler, 1968):

$$\begin{aligned} E &= -\left(\alpha + 0.01592\alpha^2\right) \hbar\omega_0, \\ m^* &= \left(1 + \alpha/6 + 0.02362\alpha^2\right) m. \end{aligned} \tag{2.4.5}$$

Hence within the accuracy of the terms $O(\alpha^3)$ the polaron energy expression calculated within the Tulub approach with the use of perturbation theory coincides with exact result (2.1.1) (see Section 2.11).

2.5 Strong coupling

The case of strong coupling is much more complicated. To reveal the character of the solution in the strong coupling region let us start with considering the analytical properties of the function $D(s)$ in the form:

$$D(s) = D(1) + \frac{s-1}{3\pi^2} \int_0^\infty \frac{k^4 f_k^2 \omega_k dk}{(\omega_k^2 - 1)(\omega_k^2 - s)}, \tag{2.5.1}$$

where $D(1)$ is the value of $D(s)$ for $s = 1$:

$$D(1) = 1 + Q \equiv 1 + \frac{1}{3\pi^2} \int_0^\infty \frac{k^4 f_k^2 \omega_k}{\omega_k^2 - 1} dk. \tag{2.5.2}$$

From (2.3.19), it also follows that:

$$D(s) = 1 - \frac{1}{3\pi^2}\int_0^\infty \frac{\omega_k k^4 f_k^2}{s - \omega_k^2}\,dk. \tag{2.5.3}$$

Function $D(s)$ being a function of a complex variable s has the following properties: **1)** $D(s)$ has a crosscut along the real axis from $s = 1$ to ∞ and has no other peculiarities; **2)** $D^*(s) = D(s^*)$; **3)** as $s \to \infty$ $sD(s)$ increases not slower than s. These properties enable us to present the function $[(s-1)D(s)]^{-1}$ in the form (Appendix C):

$$\frac{1}{(s-1)D(s)} = \frac{1}{2\pi i}\oint_{C+\rho} \frac{ds'}{(s'-s)(s'-1)D(s')}, \tag{2.5.4}$$

where contour $C + \rho$ is shown in Fig. 2.2:

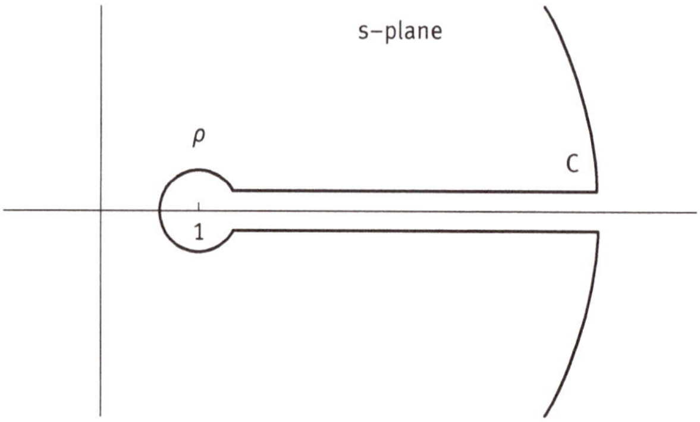

Fig. 2.2: Contour $C + \rho$.

The integrand function in (2.5.4) has a pole at $s' = 1$ and a section from $s' = 1$ to $s' = \infty$. Having performed integration in (2.5.4) along the upper and bottom sides of the crosscut we get the following integral equation for $D^{-1}(s)$:

$$\frac{1}{D(s)} = \frac{1}{1+Q} + \frac{s-1}{3\pi^2}\int_0^\infty \frac{k^4 f_k^2 \omega_k dk}{(s-\omega_k^2)(\omega_k^2 - 1)\left|D(\omega_k^2)\right|^2}. \tag{2.5.5}$$

With the use of integration by parts expression (2.3.18) for ΔE can be written as follows:

$$\Delta E = \frac{1}{2\pi^2}\int_0^\infty dk k^4 f_k^2 \omega_k \frac{1}{2\pi i}\oint_C \frac{\sqrt{s}}{(s-\omega_k^2)^2}\frac{1}{D(s)}ds. \tag{2.5.6}$$

From (2.5.5) and (2.5.6), we have

$$\Delta E = \frac{1}{2\pi^2}\int_0^\infty \frac{k^4 f_k^2}{2(1+Q)}dk + \frac{1}{12\pi^4}\int_0^\infty\int_0^\infty \frac{k^4 f_k^2 p^4 f_p^2 \omega_p(\omega_k\omega_p + \omega_k(\omega_k+\omega_p)+1)}{(\omega_k+\omega_p)^2(\omega_k^2-1)\left|D(\omega_k^2)\right|^2}dpdk. \quad (2.5.7)$$

Equation (2.4.1) for μ_k^{-1} according to (2.5.5) can be presented in the form:

$$\mu_k^{-1} = \frac{1}{1+Q} + \frac{1}{3\pi^2}\int_0^\infty \frac{p^4 f_p^2(\omega_k\omega_p+1)dp}{\left(\omega_p^2-1\right)(\omega_k+\omega_p)\left|D\left(\omega_p^2\right)\right|^2}. \quad (2.5.8)$$

Equations (2.4.1) and (2.5.8) for finding f_k as well as expressions in (2.3.20) and (2.5.7) for calculating polaron energy are very complicated and their exact solution can hardly be obtained. To calculate approximately the energy E given by (2.3.20) and (2.5.7) in Tulub (1961), a direct variational principle was used. For the probe function, the author used Gaussian function of the form:

$$f_k = -V_k \exp(-k^2/2a^2), \quad (2.5.9)$$

where a is a variable parameter, besides, as can be seen in the case of strong coupling, $a \gg 1$. Substitution of (2.5.9) into (2.3.19) yields for real and imaginary parts of $D(s)$ (see Appendix D):

$$\mathrm{Re}D(\omega_k^2) = 1 + \lambda v(y), \quad \mathrm{Im}D(\omega_k^2) = k^3 f_k^2/6\pi,$$

$$v(y) = 1 - ye^{-y^2}\int_0^y e^{t^2}dt - ye^{y^2}\int_y^\infty e^{-t^2}dt, \quad (2.5.10)$$

$$\lambda = 4\alpha a/2\sqrt{2\pi}, \quad y = k/a.$$

In the limit of strong coupling ($\alpha \gg 1$), the expression for energy E, given by (2.3.20) with the use of (2.5.7), takes on the form:

$$E = \frac{3}{16}a^2\left[1 + q\left(\frac{1}{\sqrt{2}}\right)\right] - \frac{\alpha a}{\sqrt{\pi}}\left(2 - \frac{1}{\sqrt{2}}\right), \quad (2.5.11)$$

$$q\left(\frac{1}{\lambda}\right) = \frac{2}{\sqrt{\pi}}\int_0^\infty \frac{e^{-y^2}(1-\Omega(y))dy}{(1/\lambda + v(y))^2 + \pi y^2 e^{-2y^2}/4}, \quad (2.5.12)$$

$$\Omega(y) = 2y^2\left\{(1+2y^2)ye^{y^2}\int_y^\infty e^{-t^2}dt - y^2\right\}.$$

As $\lambda \to \infty$, integral (2.5.12) has maximum for $y^4 = 3\lambda/4$, if the function f_k is chosen in the form (2.5.9), however if the actual boundedness of the region of integration with respect to y is taken into account, this peculiarity does not take place (see Section 2.12).

When calculating (2.5.12) in Tulub (1961), Tulub assumed that in the strong coupling limit $1/\lambda = 0$. As a result of numerical integration $q(0)$ was found to be $q(0) = 5.75$, whence, varying energy E (2.5.11) with respect to a we get:

$$E = -0.105\alpha^2\hbar\omega_0. \tag{2.5.13}$$

Comparison of (2.5.13) with (2.1.6) shows that the value of E obtained for $\alpha \to \infty$ lies higher than the exact value in Pekar's theory (2.1.6). For this reason, until quite recently it was believed that Tulub's theory as applied to a polaron does not give any new results.

The situation changed radically after publication of Kashirina et al. (2012). There it was shown that the choice of the wave function for minimizing energy (2.3.20) in the form (2.5.9) is not optimal since it does not satisfy virial relations. As is shown in Kashirina et al. (2012), an appropriate function f_k should contain the multiplier $\sqrt{2}$ outside the exponent in expression (2.5.9). As a result, instead of (2.5.11), the polaron energy will be:

$$E = \frac{3a^2}{16}\left[1 + q\left(\frac{1}{\lambda}\right)\right] - \sqrt{\frac{2}{\pi}}\alpha\, a \tag{2.5.14}$$

Minimization of energy (2.5.14) with the optimal probe function yields:

$$E = -0.12572\alpha^2. \tag{2.5.15}$$

Result (2.5.15) is fundamental. Above all it means that Pekar's ansatz does not give an exact solution. Though result (2.5.15) refers to a particular case of Pekar–Fröhlich Hamiltonian with V_k given by (2.2.8), the conceptual conclusion should be valid for all types of self-localized states. Of special interest is to consider the case of bipolarons (Sections 2.7 and 2.8) since they can play an important role in superconductivity.

2.6 Induced charge of translation-invariant polarons and bipolarons with broken translational symmetry

To find the charge induced by an electron in a polar medium and the polarization of the medium $\mathbf{P}(r)$, we will proceed from the fact that the electrostatic potential created by the medium $\varphi(r)$ and induced in it by the electron is determined by the operator $\hat{\varphi}(r)$:

$$\hat{\varphi}(r) = -\frac{1}{e}H_{int}(r) \tag{2.6.1}$$

$$H_{int}(r) = \sum_k V_k\left(a_k e^{ikr} + a_k^+ e^{-ikr}\right) \tag{2.6.2}$$

From (2.6.1) and (2.6.2) it follows that:

$$\hat{\varphi}(r) = -\sqrt{\frac{2\pi\omega_0}{\tilde{\varepsilon}V}}\sum_k \frac{1}{k}(a_k e^{ikr} + a_k^+ e^{-ikr}) \tag{2.6.3}$$

Relevant $\hat{\varphi}$ intensity of the electric field is determined by the relation $\hat{\mathbf{E}} = -grad\,\hat{\varphi}$. With regard to the relation $\hat{\mathbf{E}} + 4\pi\mathbf{P} = 0$ the polarization $\mathbf{P}$ will be:

$$\mathbf{P}(r) = \sqrt{\frac{\omega_0}{8\pi\tilde{\varepsilon}V}}\sum_k \frac{\mathbf{k}}{k}(a_k e^{ikr} + a_k^+ e^{-ikr}). \tag{2.6.4}$$

From (2.6.1) and (2.6.2) it also follows that the density of the induced charge determined by the Poisson equation is

$$\hat{\rho}_{ind}(r) = \Delta_r\hat{\varphi}(r)/4\pi \tag{2.6.5}$$

In theories of a polaron with spontaneously broken symmetry, it is assumed that the electron's center of gravity is localized at a point r_0. For an electron located at a point r_0 with energy $H_{int}(r_0)$, as a result of averaging over the phonon variables, from (2.6.1) the following distribution of the induced potential is obtained as a function of distance $r - r_0$:

$$\varphi(r) = \langle 0|S_2^{-1}S_1^{-1}(r)\hat{\varphi}(r_0)S_1(r)S_2|0\rangle = -\frac{2}{e}\sum_k V_k f_k \cos\mathbf{k}(\mathbf{r} - \mathbf{r}_0). \tag{2.6.6}$$

Thus, for example, in the case of weak coupling, with the use of expression (2.4.3) for f_k, the quantity $\rho_{ind}(r)$ determined by (2.6.5) (Lee et al., 1953) was expressed as follows:

$$\rho_{ind}(r) = -\frac{em\omega_0}{\hbar\tilde{\varepsilon}}\frac{1}{r}\exp(-ur), \quad u = \left(\frac{2m\omega_0}{\hbar}\right)^{1/2}, \tag{2.6.7}$$

where u^{-1} is the characteristic size of the localized state. Accordingly, localized distributions are obtained for $\varphi(r)$ and $\mathbf{P}(r)$. Relevant (2.6.7) induced charge will be equal to $Q_{ind} = \int \rho_{ind} dV = -e/\tilde{\varepsilon}$.

In the TI polaron theory, the assumption about the localization of the electron in the vicinity of the point r_0 is not made. As a result, the induced potential is

$$\varphi(r) = \langle 0|S_2^{-1}S_1^{-1}(r)\hat{\varphi}(r_0)S_1(r)S_2|0\rangle = -\frac{2}{3}\sum_k V_k f_k = \text{const}, \tag{2.6.8}$$

that is, it is independent of the electron position. Accordingly, the polarization field $\mathbf{P}(r)$, determined by (2.6.4) and induced charge density $\hat{\rho}_{ind}(r)$, determined by (2.6.5) will be equal to zero.

It is important to note here that the average number of phonons N in a polaron "cloud":

$$N_p = \langle 0|S_2^{-1}\hat{N}S_2|0\rangle = \sum_k f_k^2, \quad \hat{N} = \sum_k a_k^+ a_k$$

is not equal to zero and for f_k, corresponding to the weak coupling limit: $f_k = -V_k/(\hbar\omega_0 + \hbar^2k^2/2m)$ is: $N_p = \alpha/2$, which corresponds to the known value obtained by Lee, Low, and Pines.

In the limit of strong coupling with the use of $f_k = -\sqrt{2}V_k\exp(-k^2/2a^2)$ for N_p we get: $N_p = 0.126\,\alpha^2$, which is much less than the result of the diagrammatic quantum Monte Carlo method: $N_p = 0.22\,\alpha^2$ (Kashurnikov and Krasavin, 2010). For a bipolaron, N_{bp} is also nonzero and proportional to the square of the EPI constant: $N_{bp} \propto \alpha^2$.

2.7 Phonon interaction of electrons in the translation-invariant theory

The problem of the interaction of two electrons in a phonon field in the case of weak interaction was first considered by Cooper (1956). The fact that there are many electrons in the system does not affect the admissibility of the two-electron approximation, since due to the Pauli principle, electrons below the Fermi surface only weakly perturb the states located above the surface. The original Hamiltonian in the Cooper problem is the Fröhlich Hamiltonian, which in the case of two electrons has the form:

$$H = -\frac{\hbar^2}{2m}\Delta_{r_1} - \frac{\hbar^2}{2m}\Delta_{r_2} + \sum_k \hbar\omega_k^0 a_k^+ a_k + U(|r_1 - r_2|) + \\ + \sum_k [V_k \exp(ikr_1)a_k + V_k\exp(ikr_2)a_k + H.c.],$$

$$U(|r_1 - r_2|) = \frac{e^2}{\varepsilon_\infty |r_1 - r_2|}, \tag{2.7.1}$$

where r_1 and r_2 are coordinates of the first and second electrons, respectively, the quantity U describes Coulomb repulsion between the electrons, $H.c.$ – Hermitian complicated terms.

In the system of the center of mass Hamiltonian (2.7.1) takes on the form:

$$H = -\frac{\hbar^2}{2M_e}\Delta_{\mathbf{R}} - \frac{\hbar^2}{2\mu_e}\Delta_{\mathbf{r}} + U(r) + \sum_k \hbar\omega_k^0 a_k^+ a_k + \\ + \sum_k 2V_k \cos\frac{\mathbf{kr}}{2}[a_k\exp(i\mathbf{kR}) + H.c.],$$

$$\mathbf{R} = \frac{r_1 + r_2}{2}, \quad r = r_1 - r_2, \quad M_e = 2m, \quad \mu_e = m/2. \tag{2.7.2}$$

In what follows we will believe that $\hbar = 1$, $\omega_k^0 = 1$, $M_e = 1$ (accordingly, $\mu_e = 1/4$).

Coordinates of the center of mass **R** can be excluded from Hamiltonian (2.7.2) through Heisenberg's canonical transformation:

$$S_1 = \exp\left(-i\sum_k \mathbf{k} a_k^+ a_k\right)\mathbf{R},$$

$$\tilde{H} = S_1^{-1} H S_1 = -2\Delta_r + U(r) + \sum_k a_k^+ a_k +$$

$$+\sum_k 2V_k \cos\frac{\mathbf{kr}}{2}(a_k + a_k^+) + \frac{1}{2}\left(\sum_k \mathbf{k} a_k^+ a_k\right)^2. \tag{2.7.3}$$

Let us consider in more detail the term corresponding to the interaction of electrons with a phonon field. Applying the Lee, Low, and Pines transformation (2.2.5) to it, we obtain for the interaction energy:

$$U_{int}(r) = \langle 0|S_2^{-1}\left(2\sum_k V_k \cos\frac{\mathbf{kr}}{2}(a_k + a_k^+)\right)S_2|0\rangle = 4\sum_k V_k f_k \cos\frac{\mathbf{kr}}{2}. \tag{2.7.4}$$

Let us find an exact expression for $U_{int}(r)$ in the limit of weak and intermediate electron–phonon interaction (EPI), which was considered by Cooper. Using for this purpose the expression for f_k given by (2.2.7), for $U_{int}(r)$ we obtain:

$$U_{int}(r) = -\frac{4e^2}{\tilde{\varepsilon} r}\left(1 - e^{-r/2r_0}\right), \tag{2.7.5}$$

where

$$r_0 = (\hbar/2m\omega_0)^{1/2} \tag{2.7.6}$$

has the meaning of a characteristic size in polaron theory. Expression (2.7.5) leads to the immediate conclusion made by Cooper: the interaction between electrons is attractive, and the Schrödinger equation corresponding to potential (2.7.5) always has a discrete electronic level lying below the Fermi surface. The latter follows from the fact that in the limit $r \to \infty$ the potential of the electronic interaction has a Coulomb form, which automatically guarantees the existence of a discrete level with negative energy.

The knowledge of $U_{int}(r)$ enables one to calculate the distribution density of the charge $\rho_{ind}(r)$ induced by electrons in a polar medium:

$$U_{int}(r) = -2e\varphi_{ind}(r) \tag{2.7.7}$$

where $\varphi_{ind}(r)$ is the potential induced by electrons. Using the Laplace equation for $\rho_{ind}(r)$ we get:

$$\rho_{ind}(r) = \Delta_r \varphi_{ind}(r)/4\pi \tag{2.7.8}$$

From (2.7.5) and (2.7.8), we express $\rho_{ind}(r)$ as follows:

$$\rho_{ind} = -\frac{e}{8\pi r_0^2 r\tilde{\varepsilon}} \exp(-r/2r_0) \tag{2.7.9}$$

From (2.7.9) it follows that the density of the induced charge tends to infinity for $r = 0$. In this case, however, the total induced charge Q_{ind}:

$$Q_{ind} = \int \rho_{ind}(r) d^3 r \tag{2.7.10}$$

is finite and equals to:

$$Q_{ind} = -2e/\tilde{\varepsilon} \tag{2.7.11}$$

Thus, the nonrelativistic quantum theory of the interaction of an electron with a polar medium gives a finite value of the induced charge, in contrast to quantum electrodynamics, where the interaction of an electron with vacuum polarization leads to the value of the induced charge equal in absolute value to the initial one (the problem of zero charge; Landau et al., 1954). In a nonrelativistic quantum theory, this would correspond to: $\varepsilon_\infty = 1$, $\varepsilon_0 = \infty$ that is, a medium with nonpolarizable electron shells of ions and an infinite static dielectric constant.

When considering the formation of a bound state, Cooper did not take into account the Coulomb repulsion between electrons. When repulsion is taken into account, the total potential U_{tot} takes the form:

$$U_{tot}(r) = U_{int}(r) + U(r) \tag{2.7.12}$$

In the absence of screening $U(r) = e^2/\varepsilon_\infty r$, a discrete level in potential U_{tot} (2.7.12) exists under the condition: $3\varepsilon_0 > 4\varepsilon_\infty$. In general, an escaped expression should be used for $U(r)$.

For example, in the Thomas–Fermi approximation $U(r) = (e^2/\varepsilon_\infty r)\exp(-r/r_{TF})$, where r_{TF} is the Thomas–Fermi radius. This changes the condition for the existence of a discrete level, making this condition less stringent.

Notice that Cooper did not use expression (2.7.5). Instead, he used a simplified expression for the Fourier components of the interaction potential: $U_{int}(k) = v/V$ for $E_F \le \hbar^2 k^2/2m \le E_F + \delta$, $v = const$ and $U_{\text{int}}(k) = 0$ for other values of k, which leads to an interaction of the form: $U_{int}(r) \approx \sin(\sqrt{2\mu_e E_F}r/\hbar)/r$ and discrete level energy Δ:

$$\Delta = \delta \exp[-1/v\rho(E_F)] \tag{2.7.13}$$

which corresponds to the radius of the state $\bar{r} \approx \hbar^2 k_F/m\Delta$, where $\rho(E_F)$ is the density of the states in the vicinity of the Fermi level. By virtue of the approximations made, Cooper did not consider the question of the energy advantage of the formation of a discrete level, which is reduced to the study of the question of the bipolaron state energy.

2.8 Intermediate coupling bipolarons

To answer the question of the value of the total energy in the Cooper problem, let us return to Hamiltonian (2.7.3). It follows from formula (2.7.3) that the exact solution of the bipolaron problem is determined by a wave function $\psi(r)$, which contains only relative coordinates r and, therefore, is translation invariant.

Averaging of $\tilde{H}$ over $\psi(r)$ leads to Hamiltonian $\bar{H}$:

$$\bar{H} = \frac{1}{2}\left(\sum_k \mathbf{k} a_k^+ a_k\right)^2 + \sum_k a_k^+ a_k + \sum_k \bar{V}_k (a_k + a_k^+) + \bar{T} + \bar{U}, \tag{2.8.1}$$

$$\bar{V}_k = 2V_k \langle \Psi | \cos\frac{\mathbf{kr}}{2} | \Psi \rangle, \quad \bar{U} = \langle \Psi | U(r) | \Psi \rangle, \quad \bar{T} = -2\langle \Psi | \Delta_r | \Psi \rangle.$$

Hamiltonian (2.8.1) differs from Hamiltonian (2.2.3) by replacing the quantity V_k into $\bar{V}_k$ in (2.2.3) and adding the constants $\bar{T}$and $\bar{U}$ to the Hamiltonian. For this reason, repeating the conclusion given in Section 2.3, for the bipolaron energy E_{bp} we obtain:

$$E_{bp} = \Delta E + \sum_k \bar{V}_k f_k + \sum_k f_k^2 + \bar{T} + \bar{U}, \tag{2.8.2}$$

where ψ and f_k are found from the condition of minimum bipolaron energy with respect to ψ and f_k. Taking into account that the value of the recoil energy ΔE involved in (2.8.2) in the weak coupling limit is equal to: $\Delta E = \sum (\hbar^2 k^2/2M_e) f_k^2$ we can present f_k in the form:

$$f_k = \frac{\bar{V}_k}{\omega_k + \hbar^2 k^2/2M_e} \tag{2.8.3}$$

We will seek the minimum of energy (2.8.2) by choosing a trial wave function ψ in the Gaussian form:

$$|\Psi(r)|^2 = \left(\frac{2}{\pi l^2}\right)^{3/2} \exp(-2r^2/l^2) \tag{2.8.4}$$

where l is a variation parameter. Substituting (2.8.3) and (2.8.4) into (2.8.2) and minimizing the resulting expression with respect to l, we represent the expression for E in the form:

$$E = -\frac{1}{24}\left[\frac{16}{\sqrt{\pi}} - \frac{8}{\sqrt{2\pi}}\frac{1}{(1-\eta)}\right]^2 \alpha^2\hbar\omega_0, \tag{2.8.5}$$

$$l = 12\left(\hbar^2\tilde{\varepsilon}/me^2\right)\Big/\alpha\left[\frac{16}{\sqrt{\pi}} - \frac{8}{\sqrt{2\pi}}\frac{1}{(1-\eta)}\right], \tag{2.8.6}$$

where l has the meaning of the characteristic size of the Cooper pair. From (2.8.5) and (2.8.6) follows the condition for the existence of a discrete level (i.e., the existence of a bipolaron state) in the limit $\alpha \to 0 : \varepsilon_0 > 1.4\varepsilon_\infty$ which is close to the criterion obtained in the previous section.

Expression (2.8.5), though corresponding to a gain in energy of the Cooper pair (i.e., the bipolaron state) for $\alpha < 1,4$, nevertheless corresponds to a metastable state. The reason is that the bipolaron state (2.8.5) is not stable with respect to its decay into two individual polaron states with energy $E = -2\alpha\hbar\omega_0$, which is fulfilled in the limit $\alpha \to 0$.

2.9 Strong coupling bipolaron

In the case of a strong coupling, the recoil energy ΔE is determined by expression (2.5.7). From expression (2.8.2), one can obtain equations for determining the energy of a bipolaron, varying E_{bp} in f_k and ψ. Since solution of the equations obtained in this way presents great difficulties for actual determination of the bipolaron energy, we use a direct variational approach, assuming (Lakhno, 2013):

$$f_k = -N\bar{V}_k \exp\left(-k^2/2\mu\right), \tag{2.9.1}$$

and choosing the wave function in the form (2.8.4), where N, μ, and l are variational parameters. For $N = 1$, expression (2.9.1) reproduces the results of the work by Lakhno (2010b), and for $N = 1$, $\mu \to \infty$ – the results of the work by Lakhno (2012a).

Substitution of (2.9.1) and (2.8.4) into the expression for the total energy (2.8.2) after minimization with respect to parameter N yields the following expression for E:

$$E(x, y;\eta) = \Phi(x, y;\eta)\,\alpha^2, \tag{2.9.2}$$

$$\Phi(x, y;\eta) = \frac{6}{x^2} + \frac{20.25}{x^2 + 16y} - \frac{16\sqrt{x^2 + 16y}}{\sqrt{\pi}(x^2 + 8y)} + \frac{4\sqrt{2/\pi}}{x(1-\eta)}.$$

Here x and y are variable parameters: $x = l\alpha$, $y = \alpha^2/\mu$, $\eta = \varepsilon_\infty/\varepsilon_0$. Let us write Φ_{min} for the minimum value of the function Φ of x and y parameters. Figure 2.3 shows the dependence of Φ_{min} on the parameter η. Figure 2.4 demonstrates the dependence of x_{min}, y_{min} on the parameter η.

Figure 2.3 suggests that $E_{min}(\eta = 0) = -0.440636\alpha^2$ yields the lowest estimation of the bipolaron ground state energy as compared to all those obtained earlier by

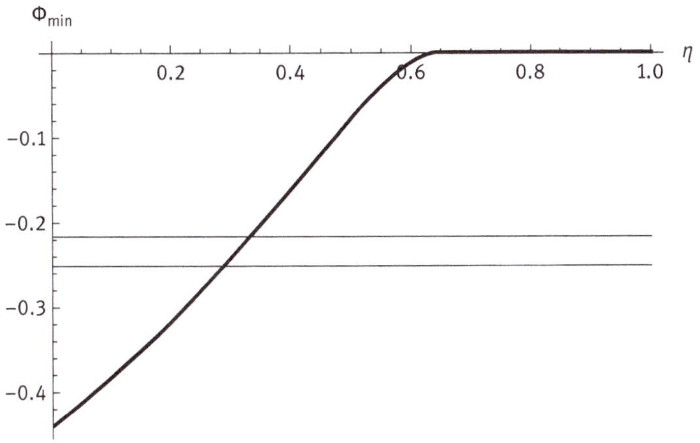

Fig. 2.3: Graph $\Phi_{min}(\eta)$.

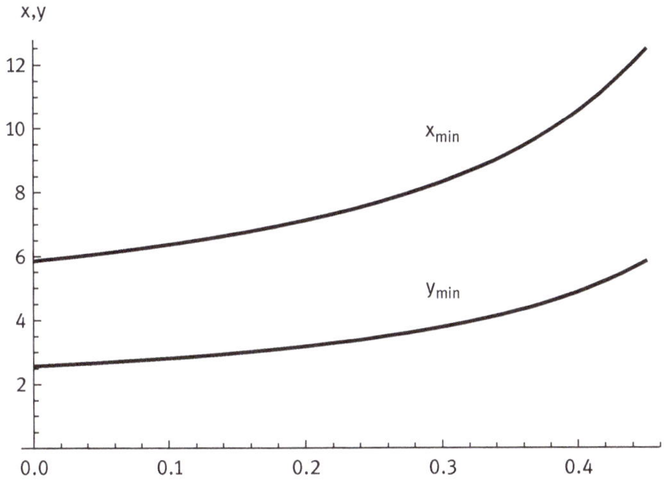

Fig. 2.4: Graphs $x_{min}(\eta)$, $y_{min}(\eta)$.

variational method (Vansant et al., 1994; Suprun and Moizhes, 1982). Horizontal lines in Fig. 2.3 correspond to the energies $E_1 = -0.217\alpha^2$ and $E_2 = -0.2515\alpha^2$, where $E_1 = 2E_{p1}$, E_{p1} is the Pekar polaron ground state energy (2.1.3); $E_2 = 2E_{p2}$, E_{p2} is the ground state energy of a TI polaron (2.5.14).

Intersection of these lines with the curve $E_{min}(\eta)$ yields the critical values of the parameters $\eta = \eta_{c1} = 0.3325$ and $\eta = \eta_{c2} = 0.289$. For $\eta > \eta_{c2}$ the bipolaron decays into two TI polarons, for $\eta > \eta_{c1}$ it breaks down into Pekar polarons. The values of minimizing parameters x_{min}, y_{min} for these values of η are: $x_{min}(0) = 5.87561$, $y_{min}(0) = 2.58537$,

$x_{min}(0.289) = 8.58537$, $y_{min}(0.289) = 3.68098$, $x_{min}(0.3325) = 8.88739$, $y_{min}(0.3325) = 4.03682$.

The critical value of the electron–phonon coupling constant α, determined from comparison of the energy expressions in the weak coupling limit (doubled energy of a weak coupling polaron: $E = -2\alpha\hbar\omega_0$) and in the strong coupling limit ($E = -0.440636\alpha^2\hbar\omega_0$), at which the translation-invariant bipolaron is formed is equal to $\alpha_c \approx 4.54$, being the lowest estimate obtained by variational method (Adamowski and Bednarek ,1992; Verbist et al., 1991, 1992; Kashirina et al., 2002, 2003, 2005; Kashirina and Lakhno, 2015). It should be emphasized that this value is conventional. Hamiltonian (2.8.1) coincides in structure with one-electron Hamiltonian (2.2.3), therefore, as in the case of a polaron, the bipolaron energy, by Gerlach and Löwen (1991), is an analytical function of α. For this reason at the point $\alpha = \alpha_c$ the bipolaron energy does not have any peculiarities and the bipolaron state exists over the whole range of α and η variation: $0 < \alpha < \infty$; $0 < \eta < 1 - 1/2\sqrt{2}$, for which $E < 0$.

To solve the problem of the existence of α_c value at which the bipolaron state can decay into individual polarons, one should perform calculations for the case of intermediate coupling. In particular, a scenario is possible when the bipolaron energy for some values of η will be lower than the energy of two individual polarons for all values of α, that is, the bipolaron state exists always. Notice, that for the derived state of the translation-invariant bipolaron, the virial theorem holds to a high precision.

The problems of arising high-temperature superconductivity (HTSC) and explaining this phenomenon by formation of bipolaron states were dealt with in a number of papers and reviews (Devreese and Alexandrov, 2009; Kashirina and Lakhno, 2010; Alexandrov and Krebs, 1992; Alexandrov and Mott, 1996; Smondyrev and Fomin, 1994). In these works the existence of HTSC is explained by Bose condensation of a bipolaron gas. The temperature of Bose condensation $T_0 = 3.31\hbar^2 n_0^{2/3}/k_B m_{bp}$, which is believed to be equal to the critical temperature of a superconducting transition T_c, for $m_{bp} \approx 10m$, depending on the bipolarons concentration n, varies in a wide range from $T_0 = 3$ K at $n = 10^{18}\text{cm}^{-3}$ to $T_0 \approx 300$ K at $n \approx 10^{21}\text{cm}^{-3}$. In the latter case the bipolarons concentration is so high that for a bipolaron gas as well as for Cooper pairs, the compound character of a bipolaron when it ceases to behave like an individual particle should show up. In the case of still higher concentrations a bipolaron should decay into two polarons. According to (2.8.4) the characteristic size of a bipolaron state is equal to l and in dimension units is written as: $l_{corr} = \hbar^2\tilde{\varepsilon}x(\eta/me^2)$. Here l_{corr} has the meaning of a correlation length. The dependence $x(\eta)$ is given by Fig. 2.4. Figure 2.4 suggests that over the whole range of η variation where the bipolaron state is stable the value of x changes only slightly: from $x(\eta = 0) \approx 6$ to $x(\eta = 0.289) \approx 8$.

Hence, even for $\eta = \eta_c$, the critical value of the concentration at which the bipolarons multipiece character is noticeable is of the order of $n_c \cong 10^{21}\text{cm}^{-3}$. This result testifies that a bipolaron mechanism of HTSC can occur in copper oxides. The theory of superconductivity on the basis of TI bipolarons will be considered in Chapter 3.

2.10 Phonon interaction in translation-invariant theory of strong coupling

In the case of a strong coupling, it is of interest to calculate the interaction between electrons as a function of the distance between them. Using (2.7.4) for the energy of the EPI (Lakhno, 2016b) for f_k, determined by (9.1) $U_{int}(r)$ has the form:

$$U_{int}(\tilde{r}) = -\sqrt{\frac{x^2+16y}{x^2+8y}}\frac{1}{\tilde{r}}F\left(\frac{2\tilde{r}}{\sqrt{16y+x^2}}\right), \qquad F(x) = \frac{2}{\sqrt{\pi}}\int_0^x e^{-t^2}dt, \tag{2.10.1}$$

where $\tilde{U}_{int}(\tilde{r}) = U_{int}(r)/\left(4me^4/\hbar^2\tilde{\varepsilon}^2\right)$ and $\tilde{r} = \left(e^2m/\hbar^2\tilde{\varepsilon}\right)r$ are dimensionless variables. The quantities $x = x(\eta)$, $y = y(\eta)$ are determined from the condition that the functional (2.8.2) be minimum.

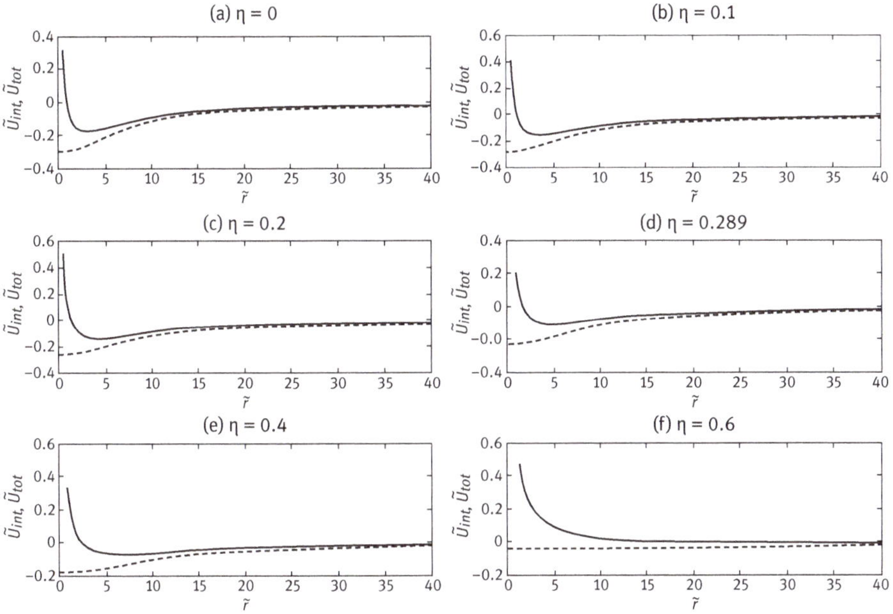

Fig. 2.5: The dependence of potential $\tilde{U}_{int}$(dashed line) and total potential $\tilde{U}_{tot}$ (straight line) on η.

Figure 2.5 shows the dependencies $\tilde{U}_{int}(\tilde{r})$ for some values of the parameter η. It can be seen that for small $\tilde{r}$ the interaction potential does not depend on $\tilde{r}$, for intermediate values the dependence is close to linear, and for large $\tilde{r}$ it has a Coulomb form: $\tilde{U}_{int}(\tilde{r}) \approx 1/\tilde{r}$. Figure 2.5 also suggests that at the point $\eta = \eta_c = 0.289$, that is, at the point where a TI bipolaron decays into TI polarons, the interaction $\tilde{U}_{int}(r)$ does not demonstrate any jumps and changes continuously as η varies up to the value of

$\eta = 1 - /2\sqrt{2}$ for which the total energy of a TI bipolaron $E_{bp} = \phi\alpha^2$ vanishes. The total interaction should also include the Coulomb interaction $U(r)$:

$$U_{tot}(r) = U_{int}(r) + U(r), \tag{2.10.2}$$

as is shown in Fig. 2.5. It looks like Coulomb interaction in the case of small r and has a near-linear shape in a certain range of r variation (this is especially clear in Fig. 2.5 (*f*): $\eta = 0.6$). This behavior reminds the interaction between quarks, with repulsive instead attractive, as in the case of quarks, Coulomb potential (polaron model of quarks was considered in (Iwao, 1976)).

The knowledge of $U_{int}(r)$ enables us to calculate the density distribution of a charge $\rho_{ind}(r)$ induced by electrons in a polar medium. Assuming:

$$U_{int}(r) = -2e\varphi_{ind}(r) \tag{2.10.3}$$

where $\varphi_{ind}(r)$ is a potential induced by the electrons, we will write for $\rho_{ind}(r)$:

$$\Delta_r \varphi_{ind}(r) = 4\pi\rho_{ind}(r) \tag{2.10.4}$$

With the use of (2.10.1), (2.10.3), and (2.10.4), we express $\rho_{ind}(r)$ as

$$\rho_{ind}(r) = \frac{32}{\pi}\sqrt{\frac{2}{\pi}\frac{e}{\varepsilon}}\left(\frac{me^2}{\hbar^2\tilde{\varepsilon}}\right)^3 \tilde{\rho}(\tilde{r}), \tag{2.10.5}$$

$$\tilde{\rho}(\tilde{r}) = \frac{1}{(x^2 + 16y)\sqrt{x^2 + y^2}} \exp\left(-\frac{8\tilde{r}}{(16y + x^2)}\right).$$

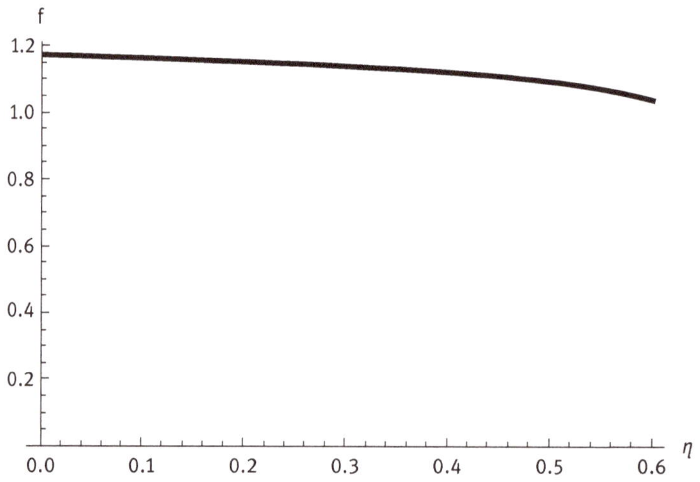

Fig. 2.6: The dependence of function $f = \sqrt{(16y + x^2)/(8y + x^2)}$ on η.

The total charge Q_{ind} induced between electrons:

$$Q_{ind} = \int \rho_{ind}(r) d^3 r \tag{2.10.6}$$

is equal to:

$$Q_{ind} = \sqrt{\frac{16y + x^2}{8y + x^2}} \frac{2e}{\tilde{\varepsilon}}. \tag{2.10.7}$$

It is pertinent to note here that the density of a charge induced in a polar medium as a function of the center of mass $\rho_{bp}(R)$ is determined by the expression:

$$\rho_{bp}(R) = -e \int \left\langle \Psi^*_{bp}(R, r) \middle| \Psi_{bp}(R, r) \right\rangle d^3 r \frac{1}{V} \tag{2.10.8}$$

where $\left|\Psi_{bp}(R, r)\right\rangle$ is given by expression (2.11.13). Thus, the density of a charge induced in a polar medium in the case of a TI bipolaron is zero. Figure 2.6 shows the dependence $f = \sqrt{(16y + x^2)/(8y + x^2)}$ as a function of the parameter η. The figure suggests that the value of a charge Q is always greater than that of $2e/\tilde{\varepsilon}$, corresponding to a bipolaron with spontaneously broken symmetry which coincide only in the limit $\eta \to 1 - 1/2\sqrt{2}$ when the distance between the polarons with broken translation symmetry is equal to infinity.

It should also be noted that the frequently used concept of interpolaron interaction in the bipolaron state in the case of TI bipolarons seems to be meaningless, since a TI bipolaron cannot be thought of as being composed of individual polarons (in this sense, the situation here is similar to quark confinement).

2.11 Ground state functional: Tulub's ansatz

To diagonalize quadratic form (2.3.4) we can use Bogolyubov–Tyablikov transformation (Tyablikov, 1967). Let us write α_k for the operators of physical particles in which H_{KB} is a diagonal operator.

Let us diagonalize the quadratic form with the use of the transformation:

$$\begin{aligned} a_k &= \sum_{k'} M_{1kk'} \alpha_{k'} + \sum_{k'} M^*_{2kk'} \alpha^+_{k'}, \\ a_k &= \sum_{k'} M^*_{1kk'} \alpha^+_{k'} + \sum_{k'} M_{2kk'} \alpha_{k'}, \end{aligned} \tag{2.11.1}$$

so that the equalities:

$$\left[a_k, a^+_{k'}\right] = \left[\alpha_k, \alpha^+_{k'}\right] = \delta_{kk'}, \quad \left[H_0, \alpha^+_k\right] = \omega_k \alpha^+_k. \tag{2.11.2}$$

be fulfilled.

Relying on the properties of unitary transformation (2.11.1) we have:

$$M_2 M_1^+ = M_1^* M_2^T,$$
$$(M_1^+)^{-1} = M_1 - M_2^* (M_1^*)^{-1} M_2. \qquad (2.11.3)$$

With the use of (2.11.3) transformation of the operators, reciprocal to (2.11.1) takes on the form:

$$\alpha_k = \sum_{k'} M^*_{1kk'} a_{k'} - \sum_{k'} M^*_{2kk'} a^+_{k'},$$
$$\alpha^+_k = \sum_{k'} M_{1kk'} a^+_{k'} - \sum_{k'} M_{2kk'} a_{k'}, \qquad (2.11.4)$$

According to Tulub (1960, 1961), matrices M_1 and M_2 take the form:

$$\left(M_{1,2}\right)_{kk'} = \frac{1}{2}\left(\omega_k \omega_{k'}\right)^{-1}\left(\omega_k \pm \omega_{k'}\right)\left[\delta(\mathbf{k}-\mathbf{k}') + (\mathbf{kk}') f_k f_{k'} \frac{2\left(\omega_k \omega_{k'}\right)^{1/2}}{\left(\omega^2_{k'} - \omega^2_k \pm i\varepsilon\right) D_\pm\left(\omega^2_k\right)}\right], \qquad (2.11.5)$$

$$D_\pm\left(\omega^2_k\right) = 1 + \frac{1}{3\pi^2}\int_0^\infty \frac{f_k^2 k^4 \omega_k}{\omega^2_k - \omega^2_p \mp i\varepsilon} dk,$$

where the superscript sign on the right-hand side of (2.11.5) refers to M_1 and the subscript sign to M_2. As a result of diagonalization quadratic form (2.3.4) changes to:

$$H_{KB} = \Delta E + \sum_k E_k \alpha^+_k \alpha_k. \qquad (2.11.6)$$

Functional of the ground state $\Lambda_0|0\rangle$ is chosen from the condition:

$$\alpha_k \Lambda_0 |0\rangle = 0. \qquad (2.11.7)$$

The explicit form of functional Λ_0 is conveniently derived if we use Fock representation for operators α_k, α^+_k (Novozhilov and Tulub, 1957, 1961) which associates operator α^+_k with some c-number $\bar{a}_k$ and operator α_k with operator $d/d\bar{a}_k$. Then, with the use of (2.11.4), condition (2.11.7) takes on the form:

$$\left(\sum_k M^*_{1kk'} \frac{d}{d\bar{a}_{k'}} - \sum_k M_{2kk'} \bar{a}_{k'}\right)\Lambda_0|0\rangle = 0. \qquad (2.11.8)$$

It is easy to verify by direct substitution in (2.11.8) that solution of equation (2.11.8) is written as:

$$\Lambda_0 = C\exp\left\{\frac{1}{2}\sum_{k,k'} a_k^+ A_{kk'} a_k^+\right\}, \tag{2.11.9}$$

where C is a constant, making operator Λ_0 unitary. To this end it is sufficient to return in (2.11.9) to quantities $\bar{a}_k$ instead of a_k^+. Hence operator Λ_0 is a squeezing operator (see Appendix E). Matrix A satisfies the conditions:

$$A = M_2^* (M_1^*)^{-1}, \quad A = A^T. \tag{2.11.10}$$

Hence, the ground state energy corresponding to functional Λ_0 is equal to:

$$\langle 0|\Lambda_0^+ H_{KB}\Lambda_0|0\rangle = \Delta E. \tag{2.11.11}$$

In Appendix A we show that $\langle 0|\Lambda_0^+ H_1\Lambda_0|0\rangle \equiv 0$.

From (2.11.9), (2.2.2), and (2.2.5), it follows that the wave function of the polaron ground state $|\psi\rangle_p$ has the form:

$$|\Psi\rangle_p = C\exp\left[-\frac{i}{\hbar}\sum_k \hbar\mathbf{k}a_k^+ a_k\mathbf{r}\right]\exp\left[\sum_k f_k(a_k^+ - a_k)\right]\Lambda_0|0\rangle. \tag{2.11.12}$$

Accordingly, the bipolaron wave function $|\Psi\rangle_{bp}$ with regard to (2.7.3) and (2.2.4) is:

$$|\Psi\rangle_{bp} = C\Psi(r)\exp\left[-\frac{i}{\hbar}\sum_k \hbar\mathbf{k}a_k^+ a_k\mathbf{R}\right]\exp\left[\sum_k f_k(a_k^+ - a_k)\right]\Lambda_0|0\rangle. \tag{2.11.13}$$

Formulae (2.11.12) and (2.11.13) imply that the wave functions of a polaron and bipolaron are delocalized over the whole space and cannot be presented as an ansatz (2.1.2).

From formulae (2.11.12) and (2.11.13), it follows that the reason why the attempt of Lee et al. (1953) to investigate the polaron ground state energy over the whole range of α variation failed was an improper choice of probe function (2.2.4) which lacks the multiplier corresponding to the functional Λ_0.

However, it should be stressed that notwithstanding a radical improvement of the wave function achieved by introducing the multiplier Λ_0 in Lee, Low, and Pines function enables one to take account of both weak and strong coupling, the results obtained by its application are not exact. The fact that Tulub's function is an ansatz follows from its properties:

$$\langle 0|\Lambda_0^+ H_1\Lambda_0|0\rangle = 0, \quad E = \langle 0|\Lambda_0^+ H_0\Lambda_0|0\rangle, \quad H_0\Lambda_0|0\rangle = E\Lambda_0|0\rangle. \tag{2.11.14}$$

Being an ansatz, Tulub's solution presents a solution of the polaron problem in a specific class of functions having the structure of $\Lambda_0|0\rangle$. That Tulub's ansatz is not an exact solution of the problem follows at least from the fact that the use of expression (2.3.20) alone for calculation of the energy, for example, in the case of weak coupling yields for E: $E = -\alpha - 1/6(1/2 - 4/3\pi)\,\alpha^2$ (Tulub, 1960). To get an exact coefficient at α^2 in the expansion of the energy in powers of α (2.1.1) we should take into account the contribution of Hamiltonian H_1, as the perturbation theory suggests (Röseler, 1968).

The fact that wave functions (2.7.12) and (2.7.13) are delocalized has a lot of important consequences, for example, the property of additivity of contributions to the energy of a polaron and a bipolaron from various branches of polarization oscillations of atoms that make up the unit cell of a crystal. The Fröhlich Hamiltonian in this case for a polaron takes the form:

$$H = -\frac{\hbar^2}{2m}\Delta_r + \sum_{j=1}^{l}\sum_{k} V_{jk}\left(a_{jk}e^{ikr} + a_{jk}^{+}e^{-ikr}\right) + \sum_{j=1}^{l}\sum_{k}\hbar\omega_{jk}^{0}a_{jk}^{+}a_{jk}, \tag{2.11.15}$$

where $j = 1, \ldots, l$ is the number of phonon branches with which an electron interacts with the force determined by the value of matrix elements V_{jk}. After removing the electron coordinates from the Hamiltonian by means of the Heisenberg transformation:

$$S_1 = \prod_{j}\exp\left(-i\sum_{j,k}\mathbf{k}a_{jk}^{+}a_{jk}\right)\mathbf{r} \tag{2.11.16}$$

instead of (2.2.3) we get a Hamiltonian:

$$\tilde{H} = \frac{1}{2m}\left(\sum_{j,k}\hbar\mathbf{k}a_{jk}^{+}a_{jk}\right)^2 + \sum_{j,k}V_{jk}\left(a_{jk} + a_{jk}^{+}\right) + \sum_{j,k}\hbar\omega_{jk}^{0}a_{jk}^{+}a_{jk}. \tag{2.11.17}$$

Hence, if it were not for the first term on the right-hand side of (2.11.17), we would have independent contributions from different phonon branches to the total Hamiltonian. Let us find the conditions under which this is fulfilled. To do this, let us see what the first term on the right-hand side of (2.11.17) transforms into after the Lee, Low, and Pines transformation:

$$S_2 = \exp\left[\sum_{j,k}f_{jk}\left(a_{jk}^{+} - a_{jk}\right)\right] \tag{2.11.18}$$

As a result of application of (2.11.18) to Hamiltonian (2.11.17), the latter takes on the form:

$$H = \sum_{j}\left(H_{j0} + H_{j1}\right) + H_{\text{int}} \tag{2.11.19}$$

where the first term on the right-hand side is the sum of the Hamiltonians of independent branches having the form (2.3.2)–(2.3.5) with v_{jk}, f_{jk}, a_{jk}, ω_{jk}^{0} instead of v_k, f_k, a_k, ω_k^0. The Hamiltonian H_{int} will include two types of terms:

$$\sum_{\substack{j,j',k,k' \\ j \neq j'}} \mathbf{k}\mathbf{k}' f_{jk}^2 f_{j'k'}^2, \tag{2.11.20}$$

$$\sum_{\substack{j,j',k,k' \\ j \neq j'}} \mathbf{k}\mathbf{k}' a_{jk}^+ a_{jk} a_{j'k'}^+ a_{j'k'}. \tag{2.11.21}$$

For even functions $f_{j,k}$, which are (2.5.9), the terms of the form (2.11.20) are equal to zero.

In calculations of the energy $E = \langle 0|\Lambda_0^+ H \Lambda_0|0\rangle$ for the wave functions $\Lambda_0|0\rangle$, where:

$$\Lambda_0 = C \exp\left(\frac{1}{2}\sum_{j,k,k'} a_{jk}^+ A_{jkk'} a_{jk'}^+\right),$$

The terms of the form (2.11.21) will vanish for the same reasons as the Hamiltonian H_1 in (2.3.2). Indeed, involved in expression:

$$\sum_{k,k'} \mathbf{k}\mathbf{k}' \langle 0|\Lambda_0^+ \sum_{j \neq j'} a_{jk}^+ a_{jk} a_{j'k'}^+ a_{j'k'}|0\rangle \tag{2.11.22}$$

mean is nothing but the norm of vector:

$$\sum_j a_{j,k}^+ a_{j,k} \Lambda_0|0\rangle$$

and therefore, is positive. Then, upon replacement $\mathbf{k} \to -\mathbf{k}$ expression (2.11.22) will change the sign and, thus, must be equal to zero. In the case of a bipolaron, similarly, it is easy to show that the bipolaron Hamiltonian H_{BP} will give the same spectrum as the Hamiltonian $\sum H_{BP,j}$ where $H_{BP,j}$ is the Hamiltonian of the bipolaron for the jth phonon branch.

For the Fröhlich Hamiltonian, in which many phonon branches appear, the quantities $V_{j,k}$ will be:

$$V_{jk} = \frac{e}{k}\sqrt{\frac{2\pi\hbar\omega_{j0}}{\tilde{\varepsilon}_j V}}, \tag{2.11.23}$$

were $\tilde{\varepsilon}_j^{-1} = \varepsilon_j^{-1} - \varepsilon_{j+1}^{-1}$, $\varepsilon_1 = \varepsilon_\infty$, $\varepsilon_{l+1} = \varepsilon_0$, $\varepsilon_1 < \varepsilon_2 < \cdots < \varepsilon_l$, $j = 1, 2, \ldots, l$. Notice that the polarization potential φ_{tot} created by all branches is equal to: $\varphi_{tot} = -\sum_{j=1}^{l} e/\varepsilon_j r = -e/\tilde{\varepsilon}$.

2.12 Discussion of the completeness of Tulub's theory

In Klimin and Devreese (2012, 2013), a question was raised as to whether Tulub's (1960, 1961) theory is complete. Arguments of Klimin and Devreese (2012, 2013) are based on the work by Porsch and Röseler (1967), which reproduces the results of Tulub's theory. However, in the last section of their paper Porsch and Röseler investigate what will happen if the infinite integration limit in Tulub's theory changes for a finite limit and then passes on to the infinite one. Surprisingly, it was found that in this case in parallel with cutting of integration to phonon wave vectors in the functional of the polaron total energy one should augment the latter with the addition δE^{PR} which input will not disappear if the upper limit tends to infinity (Porsch and Röseler, 1967; Klimin and Devreese, 2013). Relying on this result, Klimin and Devreese (2012, 2013) concluded that Tulub did not take this addition into account and therefore his theory is incomplete.

To resolve this paradox let us consider the function $\Delta(s)$ determined by formula (2.3.14) (accordingly, (2.3.19) in continuous case). As formulae (2.3.14) and (2.3.19) imply, zeros in this function contribute into "polaron recoil" energy ΔE given by (2.3.16) and, according to (2.3.15) are found from the solution of the equation:

$$1 = \frac{2}{3}\sum_k \frac{k^2 f_k^2 \omega_k}{s - \omega_k^2}. \tag{2.12.1}$$

If the cutoff is absent in the sum on the right-hand side of equation (2.12.1), then the solution of this equation yields a spectrum of s values determined by frequencies ν_{k_i} lying between neighboring values of ω_{k_i} and $\omega_{k_{i+1}}$ for all the wave vectors k_i. These frequencies determine the value of the polaron recoil energy:

$$\Delta E = \frac{1}{2}\sum_{k_i} \left(\nu_{k_i} - \omega_{k_i}\right). \tag{2.12.2}$$

Let us see what happens with the contribution of frequencies ν_{k_i} into ΔE in the region of the wave vectors k where f_k vanishes but nowhere becomes exactly zero. From (2.12.1) it follows that as $f_k \to 0$, solutions of equation (2.12.1) will tend to ω_{k_i}: $\nu_{k_i} \to \omega_{k_i}$. Accordingly, the contribution of the wave vectors region into ΔE, where $f_k \to 0$, will also tend to zero.

In particular, if we introduce a certain k^0 such that in the region $k > k^0$ the values of f_k are small, we will express ΔE in the form:

$$\Delta E = \frac{1}{2}\sum_{k_i \le k^0} \left(\nu_{k_i} - \omega_{k_i}\right), \tag{2.12.3}$$

which does not contain any additional terms. To draw a parallel with Tulub's approach, there we could put the upper limit k^0, but no additional terms would appear.

For example, if in an attempt to investigate the minimum of Tulub's functional (2.3.20), (2.5.7) we choose the probe function f_k not containing a cutoff in the form (Lakhno, 2013):

$$f_k = -V_k \exp(-k^2/2a^2(k)),$$
$$a(k) = \frac{a}{2}\left[1 + th\left(\frac{k_b - k}{a}\right)\right], \tag{2.12.4}$$

where a is a parameter of the Tulub probe function (2.5.9), k_b satisfies the condition $a \ll k_b \ll k_{oc}$, $k_{oc} = a\sqrt[4]{3\lambda/4}$ is the value of the wave vector for which Tulub's integral (2.5.12) has a maximum (Tulub, 1961; Lakhno, 2012b), then with the use of (2.12.4) in the limit $a \to \infty$, Tulub's integral $q(1/\lambda)$ will be written as

$$q\left(\frac{1}{\lambda}\right) \approx 5.75 + 6\left(\frac{a}{k_b}\right)^3 \exp\left(-\frac{k_b^2}{a^2}\right). \tag{2.12.5}$$

The second term on the right-hand side of (2.12.5) vanishes as $k_b/a \to \infty$ and we get, as we might expect, Tulub's result: $q(1/\lambda) \approx 5.75$.

Equation (2.12.1), however, has a peculiarity. Even in the case of a continuous spectrum, for $f_k = 0$, if $k > k^0$, it has an isolated solution ν_{k^0} which differs from the maximum frequency ω_{k^0} by a finite value. This isolated solution leads to an additional contribution into ΔE:

$$\Delta E = \frac{1}{2}\sum_{k_i < k^0}\left(\nu_{k_i} - \omega_{k_i}\right) + \delta E^{PR},$$
$$\delta E^{PR} = \frac{3}{2}\left(\nu_{k^0} - \omega_{k^0}\right). \tag{2.12.6}$$

where ν_{k^0} has the meaning of "plasma frequency" (Porsch and Röseler, 1967). Hence, here a continuous transition from the case of $f_k \to 0$ for $k > k^0$ to the case of $f_k = 0$ for $k > k^0$ is absent. As is shown by direct calculation (Appendix F), of the contribution of the term with "plasma frequency" δE^{PR} into (2.12.6), even for $k^0 \to \infty$, Porsch and Röseler theory does not transform itself into Tulub's theory.

In Tulub's theory we choose such f_k, which lead to the minimum of the functional of the polaron total energy. In particular, the choice of the probe function in the form (2.12.4) provides the absence of a contribution from "plasma frequency" into the total energy and in actual calculations one can choose a cutoff f_k, without introducing any additional terms in Tulub's functional (Lakhno, 2012b; Appendix J).

To sum up, critical remarks in Klimin and Devreese (2012, 2013) are inadequate. Their inadequacy was demonstrated in Lakhno (2012b, 2013) and in Appendix F reproduced here. At the present time, Tulub's theory and the results obtained on its basis (Kashirina et al., 2012; Lakhno, 2010b, 2012a, 2013) give no rise to doubt.

2.13 Consequences of the existence of translation-invariant polarons and bipolarons

According to the results obtained, the ground state of a TI polaron is a delocalized state of the electron–phonon system: the probabilities of electrons occurrence at any point of the space are similar. The explicit form of the wave function of the ground state is presented in Section 2.11. Both the electron density and the amplitudes of phonon modes (corresponding to renormalized by interaction frequencies v_{q_i}) are delocalized.

It should be noted that according to (2.3.15) renormalized phonon frequencies v_{q_i} in the case of a TI polaron have higher energies than nonrenormalized frequencies of optical phonons and, therefore, higher energies than nonrenormalized frequencies of a polaron with spontaneously broken symmetry (Lakhno and Chuev, 1995).

The concept of a polaron potential well (formed by local phonons (Lakhno and Chuev, 1995)) in which an electron is localized, that is, the self-trapped state, is lacking in the translation-invariant theory. Accordingly, the induced polarization charge of the TI polaron is equal to zero. Polarons lacking a localized "phonon environment" suggests that its effective mass is not very much different from that of an electron. The ground state energy of a TI polaron is lower than that of Pekar's polaron and is given by formula (2.5.15) (for Pekar's polaron the energy is determined by (2.1.6)).

Hence, for zero total momentum of a polaron, there is an energy gap between the TI polaron state and the Pekar one (i.e., the state with broken translation invariance). The TI polaron is a structureless particle (the results of investigations of the Pekar polaron structure are summed up in Lakhno and Chuev (1995)).

According to the TI polaron theory, the terms "large-radius polaron" (LRP) and "small-radius polaron" (SRP) are relative, since in both cases the electron state is delocalized over the crystal. The difference between the LRP and SRP in the translation-invariant theory lies in the fact that for the LRP the inequality $k_{char}a < \pi$ is fulfilled, while for the SRP $k_{char}a > \pi$ holds, where a is the lattice constant and k_{char} is a characteristic value of the phonon wave vectors making the main contribution into the polaron energy. This statement is valid not only for Pekar–Fröhlich polaron, but for the whole class of polarons whose coupling constant is independent of the electron wave vector, such as Holstein polaron, for example. For polarons whose coupling constant depends on the electron wave vector, these criteria may not hold (as is the case with Su–Schrieffer–Heeger model, for example, Marchand et al. (2010)).

These properties of TI polarons determine their physical characteristics which are qualitatively different from those of Pekar's polarons. When a crystal has minor local disruptions, the TI polaron remains delocalized. For example, in an ionic crystal containing vacancies, delocalized polaron states will form *F*-centers only at a certain critical value of the static dielectric constant ε_{0c}. For $\varepsilon_0 > \varepsilon_{0c}$ a crystal will have delocalized TI polarons and free vacancies. For $\varepsilon_0 = \varepsilon_{0c}$ a transition from the delocalized state to that localized on vacancies (collapse of the wave function) will take place. Such behavior of TI polarons is qualitatively different from that of Pekar's polarons which are localized on the vacancies at any value of ε_0. This fact accounts for, in particular, why free Pekar polaron does not demonstrate absorption (i.e., structure), since in this case the TI polaron is realized. Absorption is observed only when a bound Pekar polaron, that is, *F*-center is formed. These statements are also supported by a set of recent papers where Holstein polaron is considered (Hague et al., 2008; Mishchenko et al., 2009; Ebrahimnejad and Berciu, 2012).

Notice that the physics of only free strong-coupling polarons needs to be changed. The overwhelming majority of results on the physics of strong-coupling polarons has been obtained for bound (on vacancies or lattice disruptions) polaron states of Pekar type and do not require any revision.

Taking account of translation invariance in the case of a polaron leads to a minor change in the assessment of the ground state, however leads to qualitatively different visions of the properties of this state. In Tulub (1961), in the section devoted to scattering of a TI polaron, Tulub shows that as the constant of electron–phonon coupling increases up to a certain critical value, scattering of an electron on optical phonons turns to zero. Hence, when the coupling constants exceed a critical value a polaron becomes superconducting. Though in ionic crystals the main mechanism of electron scattering is scattering on optical phonons (Schultz, 1959), it might appear that the contribution of acoustic phonons should also be taken into account in this case. However, as it follows from the law of conservation of energy and momentum, a TI polaron will scatter on acoustic phonons only if its velocity exceeds that of sound (Kittel, 1963).

As distinct from polarons, TI bipolarons have much greater binding energy. This leads to some important physical consequences. In particular, when a crystal has minor local disruptions, a TI bipolaron will be delocalized. Thus, in an ionic crystal with lattice vacancies, formation of *F'*-centers by delocalized bipolarons will take place only at a certain critical value of the static dielectric constant ε_{0c_1}. For $\varepsilon_0 > \varepsilon_{0c_1}$ the crystal will contain delocalized TI bipolarons and free vacancies. In the case of $\varepsilon_0 = \varepsilon_{0c_1}$ TI bipolarons will pass on from the delocalized state to that localized on vacancies, that is, to *F'*-center. Such behavior of TI bipolarons is qualitatively different from the behavior of polarons with spontaneously broken symmetry of Pekar type (Kashirina and Lakhno, 2010), which are localized on the vacancies at any value of ε_0.

Delocalized for $P = 0$ TI bipolarons, where P is the total momentum of a bipolaron will be separated by an energy gap from bipolaron states with broken translation invariance which are described by a localized wave function. This problem is lacking in in the translation-invariant bipolaron theory of superconductivity (Chapter 4). It should be emphasized that the above-mentioned properties of TI bipolarons impart them superconducting properties even in the absence of their Bose condensation, and the high binding energy of bipolarons makes such a scenario of superconductivity real even in highly defective crystals. As with TI polarons, in the case when the coupling constant exceeds a certain critical value, TI bipolarons become superconducting. As is known, interpretation of the high-temperature superconductivity relying on the bipolaron mechanism of Bose condensation runs into a problem associated with a great mass of bipolarons and, consequently low temperature of Bose condensation. The possibility of smallness of TI bipolarons mass resolves this problem. It should be stressed that the above-mentioned properties of translation-invariant bipolarons impart them superconducting properties even in the absence of Bose condensation, while the great binding energy of bipolarons substantiates the superconductivity scenario even in badly defect crystals.

2.14 Quantum field theory methods and TI polarons

At present, Tulub's theory and the quantitative results obtained on its basis are beyond doubt. The considered quantum field theory is nonperturbative and can reproduce not only the limits of strong and weak coupling, but also the mode of intermediate coupling.

One of the most effective methods for calculating polarons and bipolarons in this range of coupling strengths is considered to be the path integration method (Feynman, 1972). This approach, without proper modification, is not translation-invariant, since in this method the main contribution to the energy levels is made by the classical solutions (i.e., the extrema of the exponent of the classical action, which appears in the path integral). Moreover, due to translation invariance, such solutions are not isolated stationary points, but belong to a continuous family of classical solutions obtained by acting on the original classical solution of the translation operator. Accordingly, the stationary phase approximation in a translation-invariant system is inapplicable.

In the quantum field theory, approaches based on the introduction of collective coordinates into the functional integral (Rajaraman, 1982) have been developed to reconstruct translational invariance (Rajaraman 1982), which, however, have not been used in polaron theory until now. For this reason, it is not surprising that the path integrals method used in the polaron theory leads to a result that coincides with the semiclassical theory of the strong coupling polaron (Gerlach and Löwen, 1988).

Recently, such a powerful computational method as the quantum Monte Carlo method has been developed in the polaron theory (Mishchenko, 2005; Kashurnikov and Krasavin, 2010). This method, itself being only a calculation tool, without the above modification, cannot reproduce the results of the Tulub ansatz. In the case of the Monte Carlo diagrammatic method, an obstacle to verifying the Tulub ansatz in the strong-coupling limit is the need to calculate diagrams of a very high order.

Let us sum up the results obtained. Pekar's ansatz (2.1.2) presents the initial premise about the form of the solution, which has been confirmed in the course of its numerous tests and proofs. Over the more than eighty-year history of the development of the polaron theory (if the starting point is from Landau's (1933) note, ansatz (2.1.2) has confirmed the opinion as an asymptotically exact solution of the polaron problem in the strong-coupling limit.

Tulub's ansatz (Section 2.11) presents another premise about the form of a solution, the structure of which is determined by the form of the function $\Lambda_0|0\rangle$. Within this premise, Tulub's solution is also asymptotically exact. Since the Tulub solution gives a lower energy value for the polaron, from the variational point of view, preference should be given to the Tulub's ansatz.

Thus, the polaron theory can in no way be considered a complete theory. Within the framework of the Tulub ansatz, there is a huge amount of work to revise many concepts (for example, superconductivity in Chapter 4) and provisions in condensed matter physics. The extension of the field of application of the Tulub ansatz to other branches of quantum field theory can lead to a radical revision of many results that seem to be unquestionable at present, and vice versa. For example, the inseparability of the bipolaron state in the polaron model of quarks (Iwao, 1976) (the role of phonons is played by the gluon field) provides a natural explanation for their confinement. The fundamental question of how does the local particle–field interaction leads to nonlocal one (action at a distance) is also illustrated by solution the problem of two electrons in phonon field (Sections 2.7–2.10). It was noted in Lakhno (2014) that in TI theory there is no need to use the Higgs mechanism of spontaneous symmetry breaking to obtain the masses of elementary particles.

2.15 Translation-invariant excitons

This section deals with large-radius excitons in polar crystals. It is shown that the translation-invariant description of excitons interacting with a phonon field leads to a nonzero contribution of phonons to the energy of the exciton ground state only in the case of a weak or intermediate strength of the EPI. It is concluded that self-trapped excitons cannot exist in the strong-coupling limit. The features of the absorption and emission spectra of translation-invariant excitons in a phonon field are discussed. The conditions are found under which the hydrogen-like exciton model remains valid under the conditions of EPI.

2.15.1 Introduction

The exciton theory represents an extensive chapter in modern condensed matter physics (Knox, 1963; Agranovich, 1968; Davydov, 1971; Rashba and Sturge, 1982; Veta et al., 1986). One of its sections is the theory of excitons in polar media (Kuper and Whitfield, 1963; Devreese and Peeters, 1984). As in the case of polarons, the description of free excitons in a homogeneous and isotropic polar medium should be translation-invariant (TI). This leads to numerous consequences. Being bosons, excitons, like bipolarons, are capable of forming a Bose condensate. However, experimental confirmation of this possibility was obtained quite recently (Kogan et al., 2017). A number of theories of superconductivity are also based on the participation of excitons in the formation of a Bose condensate.

The extensive literature on excitons provides detailed coverage of numerous exciton-related phenomena. For this reason, we will focus only on some of the qualitative differences that result from the translation-invariant theory of EPI from the theory of self-trapped excitons in polar media.

In most modern works on excitons, when interpreting their spectral lines, the presence of the environment, in particular, a polar medium (in the case of polar crystals) is simply ignored (see the review by Aßmann and Bayer (2020) and the literature cited therein). If the influence of the medium is taken into account, then a clear picture of the spectral lines of excitons should have been absent. Experimentally, one can clearly see distinguishable peaks corresponding to transitions to highly excited states with a very large energy number. It seems completely inexplicable how the presence of an environment and a strong EPI, which should lead to shifts and broadening of the exciton transition lines, as well as to distortions of the shape of its spectrum, leave the closely located lines of transitions to highly excited states distinguishable.

As a result of numerous theoretical studies of this issue, an idea was formed about the significant contribution of EPI into the binding energy of an exciton, expressed in the replacement of a simple hydrogen-like model with a modified one, in which consideration of the polarization cloud surrounding the electron and hole, that is, the polaron effect, is achieved by replacing the Coulomb interaction with the screened one. The most popular interaction potentials used in the interpretation of experimental observations are the potentials by Haken (1958),Tulub (1958), Bajaj (1974), and Pullmann and Büttner (1977). Nevertheless, in the overwhelming majority of works, it is more efficient to use the simplest hydrogen-like model.

We can give the following explanation for the failure of the model potentials (Haken, 1958; Tulub, 1958; Bajaj, 1974; Pullmann and Büttner, 1977). The point is that the model potentials by Haken (1958), Tulub (1958), Bajaj (1974), Pullmann and Büttner (1977) were obtained to approximate the binding energy of an exciton in the ground state, and then used to calculate the energy levels in such a potential. In fact, an approach would be correct, in which each excited state would be associated with its own (self-consistent) potential, for example, as was done in Lakhno and Balabaev (1983) for *F*-centers.

This problem, however, due to its great complexity, was not solved. As will be shown in this work, in reality, the solution of such a problem is not required, since in the case of a strong EPI and the presence of translation invariance, the exact spectrum of the exciton is hydrogen-like. This explains the success of its widespread use.

Thus, the main result is that in TI systems, self-trapped (self-consistent) states of excitons caused by EPI are impossible, just as self-trapped states of a polaron and a bipolaron are impossible (Lakhno 2010b, 2012a, 2013, 2015b; Kashirina et al., 2012). At the same time, as will be shown below, the presence of translational invariance leads to important features in the spectra of TI excitons.

2.15.2 Hamiltonian of an exciton in a polar crystal

The Hamiltonian of an exciton in a polar crystal is the Pekar–Fröhlich Hamiltonian, which describes the interaction of an electron and a hole with each other and with optical phonons:

$$\hat{H} = -\frac{\hbar^2}{2m_1}\Delta_{r_1} - \frac{\hbar^2}{2m_2}\Delta_{r_2} + \sum_k \hbar\omega_0(k)\, a_k^+ a_k - \frac{e^2}{\varepsilon_\infty|r_1 - r_2|} + \\ + \sum_k \left(V_k e^{ikr_1} a_k - V_k e^{ikr_2} a_k + H.c.\right), \qquad (2.15.1)$$

$$V_k = \frac{e}{|k|}\sqrt{\frac{2\pi\hbar\omega_0(k)}{V\tilde{\varepsilon}}}, \quad \omega_0(k) = \omega_0, \quad \tilde{\varepsilon}^{-1} = \varepsilon_\infty^{-1} - \varepsilon_0^{-1},$$

where e is the electron charge, m_1 and m_2 are masses of an electron and a hole, ε_∞ and ε_0 are high frequency and static dielectric permittivities, r_1 and r_2 are coordinates of an electron and a hole, $\omega_0(k)$ is a phonon frequency which in the case of optical phonons is independent of k and equal to ω_0.

Hamiltonian (2.15.1) corresponds to the case of a continuous polar medium, that is, the case of Wannier–Mott exciton in a polar medium. Opposite signs in the interaction Hamiltonian (2.15.1) correspond to opposite signs of the charge of an electron and a hole.

Having passed in Hamiltonian (2.15.1) from r_1 and r_2 to coordinates of the center of mass R and relative coordinates r:

$$\mathbf{r}_1 = \mathbf{R} + \frac{m_2}{M}\mathbf{r}, \quad \mathbf{r}_2 = \mathbf{R} - \frac{m_1}{M}\mathbf{r}, \quad M = m_1 + m_2, \quad \mu = \frac{m_1 m_2}{M}, \qquad (2.15.2)$$

we get:

$$\hat{H} = -\frac{\hbar^2}{2M}\Delta_{\mathbf{R}} - \frac{\hbar^2}{2\mu}\Delta_{\mathbf{r}} + \sum_k \hbar\omega_0(\mathbf{k})\, a_k^+ a_k - \frac{e^2}{\varepsilon_\infty |r|} + \tag{2.15.3}$$
$$+\sum_k V_k a_k \left[e^{i\mathbf{k}(\mathbf{R}+m_2\mathbf{r}/M)} - e^{i\mathbf{k}(\mathbf{R}-m_1\mathbf{r}/M)}\right] + H.c.$$

Having eliminated in (2.15.3) coordinates of the center of mass of an exciton via Heisenberg operator $S = \exp\left(-i/\hbar \sum \hbar kRa_k^+ a_k\right)$ and averaged the Hamiltonian obtained over the wave function of relative motion $\Psi(r)$ we obtain:

$$\hat{\hat{H}} = \frac{1}{2M}\left(\sum_k \mathbf{k} a_k^+ a_k\right)^2 + \sum_k \hbar\omega_0(k) a_k^+ a_k + \sum_k \left[\bar{V}_k a_k + H.c.\right] + \bar{T} + \bar{U}, \tag{2.15.4}$$

$$\bar{T} = -\frac{\hbar^2}{2M}\int \Psi^* \Delta_r \Psi d^3r, \quad \bar{U} = -\frac{e^2}{\varepsilon_\infty}\int \Psi^* \frac{1}{|r|}\Psi d^3r,$$

$$\bar{V}_k = V_k \langle \Psi | \exp i\mathbf{k}\mathbf{r}m_2/M - \exp(-i\mathbf{k}\mathbf{r}m_1/M)|\Psi\rangle.$$

Let us consider different limiting cases for this Hamiltonian.

2.15.3 The ground state of an exciton in a polar crystal in the case of weak and intermediate electron–phonon interaction

The contribution of EPI into the exciton energy in the case of weak coupling for $m_1 \neq m_2$ is nonzero and leads to a decrease in the exciton energy. This automatically follows from expression (2.15.4), which differs from the case of a polaron by substitutions $V_k \to \bar{V}_k$; $m_1, m_2 \to \mu, M$ and adding constants $\bar{T}$ and $\bar{U}$ to the Hamiltonian. As a

result, for the energy of the ground state of a resting exciton in the case of weak and intermediate coupling, according to Lee et al. (1953), we obtain:

$$E = \bar{T} + \bar{U} - \sum_k \frac{|\bar{V}_k|^2}{\hbar\omega_0(k) + \hbar^2 k^2/2M} \tag{2.15.5}$$

According to Gerlach and Luczak (1996), the energy of the ground state in this limit is

$$E = -(\alpha_1 + \alpha_2)\hbar\omega_0 - R_0\frac{\mu_p}{\mu}, \tag{2.15.6}$$

$$R_0 = \mu e^4/2\hbar^2\varepsilon_0^2, \quad \mu_p = \frac{m_1^p m_2^p}{m_1^p + m_2^p}, \quad m_i^p = m_i(1+\alpha_i/6),$$

$$\alpha_i = \frac{1}{2}\frac{e^2 u_i}{\hbar\omega_0 \tilde{\varepsilon}}, \quad u_i = \left(\frac{2m_i\omega_0}{\hbar}\right)^{1/2},$$

where α_i, $i = 1, 2$ have the meaning of EPI constants for an electron and a hole, respectively.

It follows from (2.15.6), that in the absence of a static Coulomb interaction between an electron and a hole ($\varepsilon_0 = \infty$), there is an ordinary polaron shift in the energies of an electron and a hole moving independently. In another limiting case, when EPI is absent: $\tilde{\varepsilon} = \infty$, from (2.15.6) follows the expression for the effective hydrogen atom in the ground state.

2.15.4 The ground state of an exciton in a polar crystal in the case of a strong electron–phonon interaction

Hamiltonian (2.15.4) does not depend on the coordinates of the center of mass of the exciton R. Hence, it follows that self-trapping of the exciton, that is, formation of an exciton localized in the R-space is impossible. This is a consequence of the fact that the total momentum of the exciton commutes with the Hamiltonian; accordingly, the eigenwave functions of the Hamiltonian are simultaneously the eigenfunctions of the operator of the total momentum **P** or plane waves in R-space.

Another important conclusion that follows from the form (2.15.4) is that in the limit of strong coupling with phonons, the exciton is not polarizable. In other words, in the strong EPI limit, the exciton behaves like a free exciton in a nonpolar medium.

Let us show this in the case when $\omega_0(k)$ does not depend on k. In this case, Hamiltonian (2.15.4) coincides in structure with the bipolaron one considered in Lakhno (2010b, 2012a, 2013, 2015b) and Kashirina et al. (2012). Repeating the calculation carried out in these works for a bipolaron in the strong-coupling limit, for the ground state energy from (2.15.4) we obtain

$$E = \Delta E + 2\sum_k \bar{V}_k f_k + \sum_k f_k^2 + \bar{T} + \bar{U}, \tag{2.15.7}$$

$$|\Psi(r)|^2 = \left(2/\pi l^2\right)^{3/2} \exp\left(-2r^2/l^2\right), \tag{2.15.8}$$

$$f_k = \pm c(V_k/\hbar\omega_0) \exp\left(-k^2/a^2\right), \tag{2.15.9}$$

where the sign «+» in (2.15.9) refers to the case $m_1 < m_2$, while the sign «-» – to the case $m_1 > m_2$; a, l, c are variational parameters involved in variational functions ψ and f_k, ΔE is the so-called recoil energy (Lakhno 2010b, 2012a, 2013, 2015b; Kashirina et al., 2012).

Substituting (2.15.8) and (2.15.9) into (2.15.7) we express the ground state energy as follows:

$$E = 0,633\frac{\hbar^2 a^2}{M} - \frac{e^2}{\sqrt{2\pi}\tilde{\varepsilon}a}\left(\frac{1}{\sqrt{\frac{l^2 m_2^2}{8M^2} + \frac{1}{a^2}}} - \frac{1}{\sqrt{\frac{l^2 m_1^2}{8M^2} + \frac{1}{a^2}}}\right)^2 + \frac{3\hbar^2}{2\mu l^2} - 2\sqrt{\frac{2}{\pi}}\frac{e^2}{\varepsilon_\infty l} \quad (2.15.10)$$

In which minimization with respect to c is already performed.

It should be noted that expression (2.15.10) gives a solution to the two-particle problem with different masses even in the case of repulsion between particles, if we change the "–" sign to "+" in the parenthesis in (2.15.10) and before the last term on the right-hand side of (2.15.10). In this case, for $m_1 = m_2$ and $a = 8/\left(\sqrt{2}l\right)$ this expression turns into the expression obtained for the bipolaron in Lakhno (2010b).

Expression (2.15.10) was obtained for the case of strong coupling, when $a \to \infty$. However, it can be shown that $E = E(l, a)$ has no minimum in this limit. The only minimum which has $E(l, a)$ corresponds to the values $l = \left(3\sqrt{\pi/8}\hbar^2\varepsilon_\infty\right)/(\mu e^2)$, $a = 0$, $E_{min} = -(4/3\pi)\mu e^4/\hbar^2\varepsilon_\infty^2$ which correspond to the case of a free exciton.

Thus, our initial assumption, made in deriving (2.15.10), about the existence of a phonon contribution into the exciton energy in the case of a strong EPI turned out to be erroneous. The result obtained indicates that the sought phonon contribution to the exciton energy, which corresponds to a polarizable exciton, can be nonzero only at finite values of a, that is, at finite values of the EPI constant.

Hence it follows that for sufficiently large values of the EPI constants of electrons and holes, when the energy of a polaron exciton is close to the energy of a free exciton, the decay of an exciton into two separate polarons with energies E_p^e and E_p^h for an electron and a hole, respectively, can become energetically more advantageous.

The condition for the stability of excitons with respect to such a decay is the fulfillment of the inequality:

$$|E^{\text{exc}}| \geq \left|E_p^h\right| + \left|E_p^e\right|. \quad (2.15.11)$$

In the case of strong coupling with the use of expressions for the energies of a free exciton $E^{\text{exc}} = -\mu e^4/2\varepsilon_\infty^2\hbar^2$ and TI polarons $E_p^{e,h} = -0.06286 m_{1,2} e^4/\tilde{\varepsilon}^2\hbar^2$ (Lakhno, 2010b, 2012a, 2013, 2015b; Kashirina et al., 2012), the exciton stability region, according to (2.15.11), will be determined by the condition:

$$0.5 - 0.5\sqrt{1 - 0.5\varepsilon_\infty^2/\tilde{\varepsilon}^2} < m_{1,2}/M < 0.5 + 0.5\sqrt{1 - 0.5\varepsilon_\infty^2/\tilde{\varepsilon}^2} \quad (2.15.12)$$

It follows from (2.15.12) that for the condition of exciton stability to be fulfilled, the static dielectric constant ε_0 must be less than $3.4\varepsilon_\infty$.

We also draw attention to the fact that functional (2.15.10) does not turn into the functional of the F-center as any of the masses tends to infinity, since such a transition would correspond to the loss of the translational invariance of the original system. As shown in Section 2.13, a free TI polaron will be captured by the Coulomb attractive charge of the F-center, only at a certain critical value of the static dielectric constant. Accordingly, in the case of free electron and hole TI polarons, such capture with the formation of an exciton state will occur only if condition (2.15.12) is satisfied.

2.15.5 Spectrum of a TI exciton

To find the spectrum of the Hamiltonian (2.15.3), we will seek a solution to problem (2.15.3) in the form:

$$\Psi = |\Psi(r)\rangle |X(R, \{a_k\})\rangle \tag{2.15.13}$$

Then, the mean over $|\Psi(r)\rangle$ value of Hamiltonian (2.15.3) will have the form:

$$\hat{\bar{H}} = \langle \Psi|\hat{H}|\Psi\rangle = -\frac{\hbar^2}{2M}\Delta_R + \sum_k \hbar\omega_0(k) a_k^+ a_k + \sum_k \bar{V}_k \left[e^{ikR} a_k + H.c.\right] + \bar{T} + \bar{U}, \tag{2.15.14}$$

which, up to constants $\bar{T}$ and $\bar{U}$ and V_k replaced by $\bar{V}_k$, defined by (2.15.4), coincides with the Hamiltonian of the polaron. Below we will set $\hbar = 1$.

Following Gerlach and Kalina (1999), we will choose the wave function $|X\rangle$, involved in (2.15.13), in the translation-invariant form:

$$|X(\mathbf{P})\rangle = \left[c_p e^{i\mathbf{PR}} + \sum_{N=1}^{\infty} \sum_{k_1,\ldots,k_N} c_{\mathbf{P},k_1,\ldots,k_N} \cdot e^{i(\mathbf{P}-k_1-k_2-\cdots-k_N)\mathbf{R}} a_{k_1}^+ a_{k_2}^+ \ldots a_{k_N}^+\right] |0\rangle \tag{2.15.15}$$

where $c_\mathbf{P}$ and $c_{\mathbf{P},k_1,\ldots,k_N}$ are normalized constants, $|0\rangle$ is a vacuum wave function, $\mathbf{P}$ is the vector of eigenvalues of the total momentum operator:

$$\hat{P} = -i\partial/\partial\mathbf{R} + \sum_{i=1}^{\infty} \mathbf{k}_i a_{k_i}^+ a_{k_i} \tag{2.15.16}$$

Since the total momentum operator (2.15.16) commutes with the Hamiltonian $\bar{H}$, the wave function $|X(\mathbf{P})\rangle$ is simultaneously their eigenfunction:

$$\begin{aligned} \hat{\bar{H}}|X(\mathbf{P})\rangle &= E(\mathbf{P})\,|X(\mathbf{P})\rangle, \\ \hat{P}|X(\mathbf{P})\rangle &= P\,|X(\mathbf{P})\rangle. \end{aligned} \tag{2.15.17}$$

Let $|X(\mathbf{P})\rangle$ be the wave function of the ground state. Then, according to Gerlach and Kalina (1999), the wave function of one-phonon excited state $\left|\Psi(\mathbf{K}_j)\right\rangle$:

$$\left|\Psi(\mathbf{K}_j)\right\rangle = a_{k_j}^{+}\left|X(\mathbf{P})\right\rangle \tag{2.15.18}$$

where $\mathbf{K}_j$ has the meaning of a total momentum in the jth excited state, will have the properties:

$$\hat{\mathbf{P}}\left|\Psi(\mathbf{K_j})\right\rangle = \mathbf{K_j}\left|\Psi(\mathbf{K_j})\right\rangle = (\mathbf{P}+\mathbf{k_j})\left|\Psi(\mathbf{K_j})\right\rangle \tag{2.15.19}$$

$$\hat{\hat{H}}\left|\Psi(\mathbf{K}_j)\right\rangle = \varepsilon(\mathbf{K}_j)\left|\Psi(\mathbf{K}_j)\right\rangle = \left(E(\mathbf{P}) + \omega_{k_i}\right)\left|\Psi(\mathbf{K}_j)\right\rangle = \left(E(\mathbf{K}-\mathbf{k}_j) + \omega_{k_j}\right)\left|\Psi(\mathbf{K}_j)\right\rangle$$

Hence, the spectrum of excited states has the form:

$$\varepsilon(\mathbf{K}) = E(\mathbf{K}_j - \mathbf{k}_j) + \omega_0(\mathbf{k}_j), \quad \omega_0(\mathbf{k}_j) = \omega_{k_j} \tag{2.15.20}$$

As for $E(\mathbf{K}_j - \mathbf{k}_j)$ in Gerlach and Kalina (1999) it was shown:

$$E(\mathbf{K}_j - \mathbf{k}_j) \le E(0) + (\mathbf{K}_j - \mathbf{k}_j)^2 / 2M \tag{2.15.21}$$

In fact, according to Lakhno (2018), in this case, instead of the inequality in (2.15.21), the exact equality holds, and for $\mathbf{K}_j = 0$ the spectrum has the form:

$$\varepsilon(\mathbf{k}_j) = E(0) + \omega_0(\mathbf{k}_j) + \mathbf{k}_j^2/2M \tag{2.15.22}$$

It should be noted that, in the general case, the wave function of an excited state containing N phonons has the form:

$$\left|\Psi_{k_1,\ldots,k_N}\right\rangle = a_{k_1}^{+}a_{k_2}^{+}\ldots a_{k_N}^{+}\left|X(\mathbf{P})\right\rangle \tag{2.15.23}$$

for which the inequality:

$$\varepsilon(k_1,\ldots,k_N) \le E(0) + \sum_{i=1}^{N}\omega_0(\mathbf{k}_j) + (\mathbf{K} - \mathbf{k}_1 - \cdots - \mathbf{k}_N)^2 / 2M \tag{2.15.24}$$

holds, where $\mathbf{K}$ is the total momentum corresponding to N-phonon excitations.

We also note that in the case of an exciton, when there is a set of electronic excitations numbered by the subscript n (by n we can mean a set of quantum numbers), instead of (2.15.22) we get:

$$\varepsilon_1(k=0) = E_1(0) = E^{exc} \tag{2.15.25}$$

$$\varepsilon_n(k \ne 0) = E_n(0) + \omega_0(\mathbf{k}) + \mathbf{k}^2/2M, \quad n = 1, 2, \ldots$$

2.15.6 Peculiarities of absorption and emission of light by TI excitons

Let us consider the case of optical phonons, when $\omega_0(\mathbf{k})$ does not depend on $\mathbf{k}$, that is, the case of polar crystals. For direct excitons, according to (2.15.25), in addition to the ordinary discrete spectrum $E_n(0)$, there is a quasi-continuous spectrum with energies $E_n(0) + \omega_0 + k^2/2M$, which makes the spectrum $E_n(0)$ distinguishable only under the condition $\omega_0 > |E_1(0) - E_2(0)|$. When the condition is met:

$$|\varepsilon_1 - \varepsilon_{n_c+1}| > \omega_0 > |\varepsilon_1 - \varepsilon_{n_c}| \tag{2.15.26}$$

only n_c first levels of an exciton will be discernible. This result can be used to study soft phonon modes associated with structural phase transitions in crystals, for example, in cuprate superconductors. Thus, if, far from the phase transition, condition (2.15.26) is fulfilled for $n_c = 2$ then the optical transition of the exciton from the ground state to the first excited state will contribute to absorption. At the point of the phase transition, when $\omega_0 \approx 0$ this contribution will be absent, since all discrete exciton levels fall into the quasi-continuous spectrum.

As is shown in Lakhno (2010b, 2012a, 2013, 2015b and Kashirina et al. (2012), the excitation spectrum (2.15.25) can be interpreted as the spectrum of renormalized EPI phonons, which represent the initial phonon with which an electron and a hole are bound. The scattering of light with a frequency ν by such phonons will lead to the appearance of satellite frequencies $\nu^{exc}_{n,k,+} = \nu + |\varepsilon_n(k)|$ and $\nu^{exc}_{n,k,-} = \nu - |\varepsilon_n(k)|$ in the scattered light. Hence, depending on the parameters involved in these expressions, a more complex structure of the absorption and emission spectrum of a TI exciton is possible in comparison with the spectrum of a free exciton. Thus, for example, when condition (2.15.26) with $n = 2$ is fulfilled, absorption (emission) of light is possible without changing the principal quantum number of the exciton n. In this case, the absorption (emission) curve will have a characteristic two-humped intensity distribution with a maximum for $\nu^{exc}_{n,k,\pm} \approx \nu \pm \omega_0$ (Snoke and Kavoulakis, 2014).

Like bipolarons, TI excitons, being bosons, can undergo Bose condensation, which was predicted in Keldysh and Kopaev (1964) and Keldysh and Kozlov (1967). In contrast to the bipolaron Bose gas, to which the statistically equilibrium description is applicable, for an exciton gas in a quasi-equilibrium photoexcited state, such a description can be applicable only for long-lived exciton states, which can be realized in semimetals, gapless semiconductors, nanodot systems, or in indirect semiconductors.

In Kogan et al. (2017), exciton condensation was probably observed in the semimetallic compound $TiSe_2$. Since the TI exciton Hamiltonian (2.15.4) is similar to the TI bipolaron one, all the results obtained in the statistical description of the TI bipolaron gas are applicable to the case of the TI exciton gas. In particular, for the temperature of Bose condensation of the TI exciton gas, we obtain (Lakhno, 2018):

$$T_c(\omega_0) = \left(F_{3/2}(0)/F_{3/2}(\omega_0/T_c)\right)^{2/3} T_c(0) \tag{2.15.27}$$

$$T_c(0) = 3,31\hbar^2 n_{exc}^{2/3}/M, \qquad F_{3/2}(x) = \frac{2}{\sqrt{\pi}} \int_0^\infty \frac{t^{1/2}dt}{e^{t+x}-1},$$

where n_{exc} is the concentration of TI excitons. Accordingly, the phase transition to the Bose-condensate exciton phase should be of the second order with a jump in the specific heat during the transition.

2.16 Conclusion

This Chapter provides an answer to the fundamental question about the role of polaron effects in the physics of excitons. Despite the fact that the important role of EPI for excitons in polar media has been established in a large number of experiments, the question of why under these conditions the hydrogen-like model is valid until now has remained open (Baranowski and Plochocka, 2020). In this Chapter, it is shown that in the case of the EPI described by the Fröhlich Hamiltonian, the hydrogen-like model turns out to be applicable if the energy of the transition to the excited state does not exceed the energy of the optical phonon.

The Pekar–Fröhlich polaron model is an indispensable component of a wide range of problems associated with the description of the properties of a particle interacting with a bosonic reservoir. Originally introduced to describe the behavior of electrons interacting with phonons in crystals, this model has found application in such diverse areas as highly correlated electronic systems, quantum information, and high energy physics. Recently, it has been actively used to describe impurity atoms in boson condensates. The results obtained in this work, in particular, explain the clearly distinguishable structure of highly excited (Rydberg) atoms surrounded by a Bose condensate (Camargo et al., 2018).

In conclusion, it should be noted that both in the theory of polarons and in the theory of an exciton interacting with phonons, there is a widespread opinion about the possibility of self-trapped polaron or exciton states. For example, by analogy with a polaron, self-trapped exciton states were considered in Dykman and Pekar (1988, 1953, 1952), Pekar et al. (1979), Sumi (1977), Shimamura and Matsura (1983), and Song and Williams (1996). They assumed that when the EPI constants exceed a certain critical value, the exciton is captured by the self-consistent potential it creates, leading to possible annihilation of an electron and hole and disappearance of the exciton. It was also believed that in the case of a very strong EPI, the energy of the lattice deformed by an exciton can exceed the energy of excitons in a rigid lattice. Changes in the energy of such deformed excitons, being negative with respect to excitons in a rigid lattice, can make advantageous spontaneous formation of excitons in crystals with a small gap, for example, in gapless polar semiconductors (excitonic matter: Veta et al., 1986; Haken, 1973).

The results obtained here exclude the possibility of the formation of self-trapped exciton states in translation-invariant systems. Conclusions about the possibility of self-trapped excitons in them are based either on a poor choice of trial variational functions, or on erroneous calculations with the use of such functions.

3 Large-radius Holstein polaron and the problem of spontaneous symmetry breaking

A translation-invariant theory is developed for a Holstein large-radius polaron whose energy is lower than that found by Holstein. The wave function corresponding to this solution is delocalized. A conclusion is made about the absence of a spontaneous symmetry breaking in the quantum system discussed.

Based on the Holstein–Hubbard model, a possibility is considered of the formation of bipolaron states in discrete molecular chains. A phase diagram determining the stability of bipolaron states in such chains is calculated.

3.1 Introduction

In the previous chapters, we considered the problem of whether the problem solution should have the same symmetry as the Hamiltonian. In this chapter, we will deal with a one-dimensional case in view of its actuality. In a classical one-dimensional lattice (molecular chain), a Bloch electron with translational symmetry will always lose its original symmetry, if we assume the possibility of deformation of such a lattice by an electron.

This problem in the case of a one-dimensional molecular crystal was first considered in the work of Holstein (1959a, 1959b). If the atoms of the lattice are considered quantum-mechanically, this conclusion will not be valid any longer. In a quantum lattice, the symmetry of the electron–phonon system is conserved if the interaction of an electron with the lattice determined by the interaction constant g is not too strong. For the value of g exceeding some critical value, according to Holstein (1959a), the symmetry is broken and a self-trapped state is formed. The statement made in Holstein (1959a) that in the strong coupling limit a self-trapped polaron state is bound to arise contradicts, however, to the fact that the total momentum of the electron–phonon system in an ideal translationally symmetrical chain should be conserved. Since the total momentum of the system is commuted with the Hamiltonian, the operator of the momentum and the operator of the Hamiltonian should have the same set of eigenfunctions. However, the eigenfunctions of the total momentum operator are plane waves, that is, delocalized states, while those of the Hamiltonian operator in the strong coupling limit are localized wave functions of the self-trapped state. This contradiction was analyzed in Gerlach and Löwen (1988, 1991) and Löwen (1988) where it was shown that for all the values of the coupling constant, the states should be delocalized. In the case of Pekar–Fröhlich polaron, these solutions in the strong coupling limit were studied in the previous chapter. In particular, it was shown that in the strong coupling limit delocalized polaron states have a lower energy than localized ones which break the symmetry.

https://doi.org/10.1515/9783110786668-003

In this chapter, we apply the approach by Tulub (1961) to the problem of a large-radius strong-coupling Holstein (1959a) polaron. We will show that in this case, as in the case of Pekar–Fröhlich polaron, in the limit of large g, minimum is reached in the class of delocalized wave functions.

3.2 Holstein polaron in the strong coupling limit: broken translation invariance

According to Lakhno (2006), Holstein Hamiltonian in a one-dimensional molecular chain in a continuum limit has the form (Appendix G):

$$\hat{H} = -\frac{\hbar^2\nabla^2}{2m} + \sum_k V_k\left(a_k e^{ikx} + a_k^+ e^{-ikx}\right) + \sum_k \hbar\omega_k^0 a_k^+ a_k;$$

$$V_k = \frac{g}{\sqrt{N}}, \qquad \omega_k^0 = \omega_0 \tag{3.2.1}$$

where a_k^+, a_k are the phonon field operators, m is an electron effective mass, ω_0 is the optical phonons frequency, and N is the number of atoms in the chain.

For the following analysis, it is convenient to present some results concerning Hamiltonian (3.2.1) in the strong coupling limit. In Holstein's (1959a) theory, as well as in Pekar's (1954) theory, it is believed that the wave function of the ground state has the form:

$$\Psi = \psi(x)\,\Phi(q_1, \ldots, q_k, \ldots) \tag{3.2.2}$$

where $\psi(x)$ is the electron wave function independent of phonon variables and Φ is the phonon wave function. The ground-state energy is determined from the condition of the total energy minimum E:

$$E = T - \Pi, \quad T = \frac{1}{2m}\int |\nabla\psi|^2 dx, \quad \Pi = \frac{g^2 a}{\hbar\omega_0}\int |\psi|^4 dx, \tag{3.2.3}$$

where a is the lattice constant. Let us introduce a scaled transformation of the wave function $\psi(x)$, retaining its normalization:

$$\psi(x) = |\xi|^{1/2}\tilde{\psi}(|\xi|x). \tag{3.2.4}$$

As a result, with the use of (3.2.3), we rewrite (3.2.2) as follows:

$$E(\xi) = |\xi|^2 T - |\xi|\,\Pi. \tag{3.2.5}$$

Figure 3.1 shows the dependence $E(\xi)$.

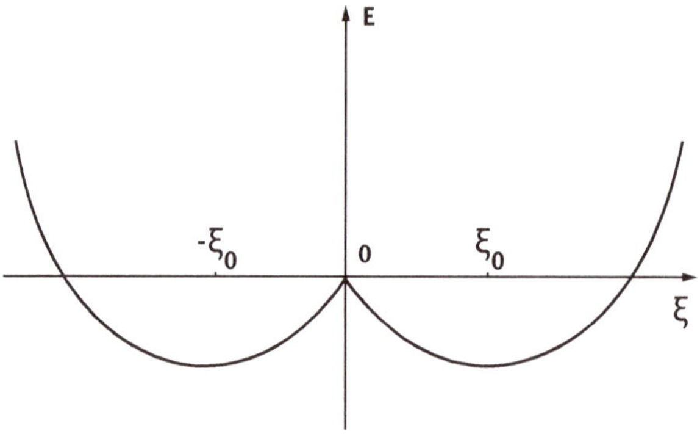

Fig. 3.1: Dependence of $E(\xi)$ on parameter ξ.

Figure 3.1 illustrates why the global symmetry of the initially symmetrical delocalized state spontaneously breaks. The reason is that this state ($\xi = 0$) corresponds to the local maximum of the functional E. The minimum of the functional at the points $\pm\xi_0$ corresponds to the energy and wave function $\psi(x)$:

$$E(\pm\xi_0) = -\frac{1}{6}\frac{\hbar^2}{mr^2}, \quad \psi(x) = \pm\left(\sqrt{2r}ch\frac{x-x_0}{r}\right)^{-1}, \quad r = \frac{\hbar^3\omega_0}{mg^2a}, \tag{3.2.6}$$

where x_0 corresponds to the position of the polaron well center in the energy minimum.

In the vicinity of this minimum, one can carry out quantizing, thereby restoring the broken symmetry. Upon this restoration a Goldstone boson (zero phonon mode) automatically arises.

A similar approach is realized in models of elementary particles (Svartholm, 1969). For example, in the standard model, not global (as in the case just considered), but local symmetry of gauge fields spontaneously breaks. In this case, Goldstone bosons do not arise, while fields become massive.

In all the cases, in these models the symmetry spontaneously breaks and the restored one turns out to be less than the initial symmetry. Below, using Holstein model as an example, we will show that the approach discussed can lead to erroneous results.

3.3 Translation-invariant theory

In the previous section, the results of the strong coupling theory with broken symmetry were given. Here, we present the general translation-invariant approach reproducing the results of Tulub (1961) as applied to Holstein Hamiltonian.

In order to make the description translation-invariant, let us exclude the electron coordinates from Hamiltonian (3.2.1) using Heisenberg transformation (2.2.2) and use Lee, Low, and Pines transformation (2.3.1) from Chapter 2. Repeating calculations made in Section 2.3, we rewrite (2.3.1) of Chapter 2 as follows:

$$\det\left|\left(s-\omega_k^2\right)\delta_{kk'}-z_k z_{k'}\right| = \prod_k \left(s-\omega_k^2\right)\left(1-\sum_k \frac{z_{k'}^2}{s-\omega_{k'}^2}\right). \tag{3.3.1}$$

Accordingly, $\Delta(s)$ is equal to

$$\Delta(s) = \left(1-\sum_k \frac{z_{k'}^2}{s-\omega_{k'}^2}\right). \tag{3.3.2}$$

As a result, for $D(s)$ instead of (2.3.19) we get:

$$\Delta E = \frac{1}{2}\sum_k \left(v_k-\omega_k\right). \tag{3.3.3}$$

$$\Delta(s) = D(s), \quad D(s) = 1-\frac{1}{\pi}\int_{-\infty}^{\infty} \frac{k^2 f_k^2 \omega_k}{s-\omega_k^2}\,dk. \tag{3.3.4}$$

With the use of (3.3.4) in the particular case of $\mathbf{P}$=0, we get the expression for the ground-state energy of operator H_0:

$$E = -\frac{1}{8\pi i}\oint_C \frac{ds}{\sqrt{s}}\ln D(s) + 2\sum_k V_k f_k + \sum_k f_k^2. \tag{3.3.5}$$

Expression for the total energy E, determined by (3.3.5) is applicable for the whole range of changes of the coupling constant g. In the next section, we will consider the case of weak coupling.

3.4 The case of weak coupling

The quantities f_k in the total energy expression E should be found from the minimum condition: $\delta E/\delta f_k = 0$, which leads to the following integral equation for finding f_k:

$$f_k = -\frac{V_k}{\left(1+k^2/2\mu_k\right)}, \quad \mu_k^{-1} = \frac{\omega_k}{2\pi i}\oint_C \frac{ds}{\sqrt{s}}\frac{1}{\left(s-\omega_k^2\right)D(s)}. \tag{3.4.1}$$

In the case of weak coupling equations (3.4.1) can be solved using the perturbation theory. In the first approximation for $g \to 0$ $D(s) = 1$, the quantity μ_k^{-1} is equal to:

$$\mu_k^{-1} = \frac{\omega_k}{2\pi i}\oint_C \frac{ds}{\sqrt{s}}\frac{1}{(s-\omega_k^2)} = 1. \tag{3.4.2}$$

Accordingly, from (3.4.1) f_k will be expressed as follows:

$$f_k = -\frac{V_k}{(1+k^2/2)}. \tag{3.4.3}$$

Hence, in the first approximation on g:

$$E = \Delta E + 2\sum_k V_k f_k + \sum_k f_k^2$$

$$\Delta E = -\frac{1}{8\pi i}\oint_C \frac{ds}{\sqrt{s}}\ln D(s)$$

$$\ln D(s) \approx -\frac{1}{\pi}\int_{-\infty}^{\infty} \frac{k^2 f_k^2 \omega_k}{s-\omega_k^2}dk. \tag{3.4.4}$$

Passing on in (3.4.4) from summation to integration with the use of the formula: $\sum_k = \frac{1}{2\pi}\int dk$ and substituting (3.4.3) into (3.4.4) with regard to expression (3.2.1) for V_k, we transform (3.4.4) into a well-known expression for the electron energy in the weak coupling limit:

$$E = -g^2\sqrt{ma^2/2\hbar^3\omega_0}, \tag{3.4.5}$$

which we write in dimensional units.

To calculate the next terms of the energy expansion in terms of the degrees of coupling constant g, one can use the approach developed in Tulub (1961) and Porsch and Röseler (1967).

3.5 The case of strong coupling

Now, we pass on to the case of strong coupling. To clear up the character of a solution in this region, let us turn to analytical properties of the function $D(s)$ (3.3.4). For this purpose, we present function $D(s)$ in the form:

$$D(s) = D(1) + \frac{s-1}{\pi}\int_{-\infty}^{\infty} \frac{k^2 f_k^2 \omega_k}{(\omega_k^2-1)(\omega_k^2-s)}dk \tag{3.5.1}$$

where $D(1)$ is the value of $D(s)$ for $s=1$:

$$D(1)=1+\frac{1}{\pi}\int_{-\infty}^{\infty}\frac{k^2f_k^2\omega_k}{\omega_k^2-1}dk\equiv 1+Q. \tag{3.5.2}$$

From (3.5.1) and (3.5.2), it follows that $D^{-1}(s)$ satisfies the integral equation:

$$\frac{1}{D(s)}=\frac{1}{1+Q}+\frac{s-1}{\pi}\int_0^{\infty}\frac{k^2f_k^2\omega_k}{(s-\omega_k^2)(\omega_k^2-1)|D(\omega_k^2)|^2}dk \tag{3.5.3}$$

whence we get:

$$\Delta E=\frac{1}{4\pi}\int_{-\infty}^{\infty}\frac{k^2f_k^2\omega_k dk}{2\,(1+Q)}+\frac{1}{4\pi^2}\int\int_{-\infty}^{\infty}\frac{k^2f_k^2p^2f_p^2\omega_p(\omega_k\omega_p+\omega_k(\omega_k+\omega_p)+1)}{(\omega_k+\omega_p)^2(\omega_p^2-1)\left|D_+(\omega_p^2)\right|^2}dpdk$$

$$D_\pm\left(\omega_p^2\right)=1+\frac{1}{\pi}\int_{-\infty}^{\infty}\frac{k^2f_k^2\omega_k dk}{\omega_k^2-\omega_p^2\pm i\varepsilon}. \tag{3.5.4}$$

To calculate the polaron energy in the strong coupling limit, let us choose the probe function f_k in the form:

$$f_k=Nge^{-k^2/2a^2}, \tag{3.5.5}$$

where N and a are variational parameters. As a result, ΔE will be written as follows:

$$\Delta E=\frac{a^2}{32}(1+2q_T), \tag{3.5.6}$$

where q_T is Tulub's (1961) integral:

$$q_T=\frac{2}{\sqrt{\pi}}\int_0^{\infty}\frac{e^{-y^2}(1-\Omega(y))dy}{v^2(y)+\pi y^2e^{-2y^2}/4}$$

$$\Omega(y)=2y^2\left\{(1+2y^2)ye^{y^2}\int_y^{\infty}e^{-t^2}dt-y^2\right\}$$

$$v(y)=1-ye^{-y^2}\int_0^{y}e^{t^2}dt-ye^{y^2}\int_y^{\infty}e^{-t^2}dt, \tag{3.5.7}$$

for which an approximate value: $q_T\approx 5.75$ was obtained in Tulub (1961) with a high degree of accuracy.

With the use of (3.3.5), (3.5.6), and (3.5.5), the ground-state energy takes the form:

$$E = \Delta E + 2\sum_k V_k f_k + \sum_k f_k^2 \approx \frac{12.5}{32} a^2 - \sqrt{\frac{2}{\pi}} \left(1 - \frac{N}{2\sqrt{2}}\right) g^2 a\, N \tag{3.5.8}$$

Minimization of the ground-state energy by parameters a and N yields the following value of the polaron ground-state energy:

$$E \approx -0.2037 \frac{m a_0^2}{\hbar^2} \frac{g^4}{\hbar^2 \omega_0^2}, \tag{3.5.9}$$

which is presented in dimensional units. This result is fundamental since the energy value obtained is lower than that of Holstein polaron (3.2.6) (Appendix E).

For the probe function chosen in the form (3.5.5), the virial theorem holds: $2T = \Pi$, $W = 3E$, where W is the electron energy. Notice that the mere fact that the virial theorem holds says nothing on whether the symmetry in the state under consideration is broken or not and for Holstein polaron, the virial theorem holds true both in the case of broken symmetry (Section 3.2), and in the state with translation symmetry discussed here.

Equating the values of the weak coupling energy (3.4.5) and the strong coupling one (3.5.9), we can, as is customary in the polaron theory, get the value of the dimensionless coupling constant $g_c = g/\hbar\omega_0$ at which a transition from weak coupling to strong one occurs:

$$g_c \approx 1.86 \sqrt[4]{\hbar / m a_0^2 \omega_0}. \tag{3.5.10}$$

It should be stressed, however, that no jump-like transition from weak and intermediate coupling to the strong one occurs. The polaron state remains delocalized for all the values of the coupling constant and $E(g)$ is an analytical function g (Gerlach and Löwen, 1988; Löwen, 1988; Korepin and Essler, 1994). This conclusion automatically results from the analytical properties of the function $D(s)$ too.

3.6 Holstein TI polaron and the problem of quantization in the neighborhood classic solution

Returning to the original Landau hypothesis that an electron wave function loses symmetry as a result of formation of a self-trapped localized state by the electron, it can be stated that it is erroneous. This can already be seen from Hamiltonian (3.2.1), which after the Heisenberg canonical transformation (2.2.2) in Chapter 2 does not contain electronic coordinates at all. Hence, it follows the general form of the solution of the Schrödinger equation for Hamiltonian (3.2.1) is

$$\Psi = \exp\left\{\frac{i}{\hbar}\left(P - \sum_k \hbar k a_k^+ a_k\right)x\right\}\Lambda_0\{a_k\}|0\rangle \tag{3.6.1}$$

where $\Lambda_0\{a_k\}$ is the functional of the field operators and $|0\rangle$ is a vacuum wave function, represents plane waves. It is important that the electronic and phonon variables, in contrast to the case of broken symmetry (Section 3.2), are not separated in (3.6.1).

According to Fig. 3.1, from the classical point of view, such a solution is unstable: any infinitesimal change in the classical orbit at this point will lead to an increase in the amplitudes of the deviation of the electron trajectory, leading to their finite values. The quantum mechanical consideration expands the space of admissible states and leads to the possibility of stable oscillations near a classically unstable point $\xi = 0$.

The quantum mechanical states determined by solution (3.6.1) lack their classical analogue. In particular, it follows from (3.6.1) that for the found solution the concept of a classical polaron well localized in space is absent, since a plane wave cannot maintain a finite number of displacements of atoms from their equilibrium positions.

The inapplicability of the adiabatic approximation (3.2.2) in translation-invariant systems can also be qualitatively illustrated by the following reasoning. The criterion for the applicability of the adiabatic approximation is the smallness of the ratio to m/M where M is the mass of the atom in the lattice. When $m/M \to 0$, that is, when the mass of an atom tends to infinity, it can be viewed as a classical particle. Accordingly, the displacement field can be regarded as classical. Then the separation of motions determined by (3.2.2) becomes physically obvious: a localized electron, described by the wave function $\psi(x)$ performing a finite motion, rapidly oscillates in a potential well, and heavy atoms perceive only its average motion, not having time to adapt at each moment to the position of the electron in space. In other words, the electron is represented as a static charge, distributed with a density $|\psi(x)|^2$.

The physical situation changes if the electron is delocalized. In this case, it makes an infinite motion, enabling even an infinitely heavy atom to shift by a finite value by the infinite time when electron approaches this atom. Thus, the adiabatic approximation (3.2.2) turns out to be invalid in this case. Accordingly, the concept of the displacement field as a classical one becomes inapplicable.

The fact that the delocalized solution has a lower energy compared to the localized one leads to numerous physical consequences similar to those discussed in the previous chapter in the case of the Pekar–Fröhlich polaron. First of all, by analogy with an electron in an ideal rigid lattice, in which Bloch electrons are superconducting, in a deformed lattice at zero temperature, delocalized polarons described by wave function (3.6.1) will be superconducting. Any attempts to divide the electron–phonon system into a polaron and a phonon field in which the polaron moves in order to calculate its mobility (Melnikov and Volovik, 1974), by virtue of (3.6.1), turns out to be impossible for a translation-invariant polaron. In the presence of defects in the lattice, translation-invariant polarons will not be captured by them if the energy gain due to the localization of the polaron on the defect is not greater than the energy gain in the formation of a translation-invariant polaron. This is the qualitative difference between translation-invariant polarons and Holstein polarons with spontaneously broken symmetry, since the latter will be localized at the defect for an arbitrarily small value of the trap potential.

Since the global symmetry for translation-invariant polarons is conserved, Goldstone modes will be absent in their spectrum, while for the Holstein polaron the zero mode will always be present in the phonon spectrum. We also note that in the case of translation-invariant polarons, the phonon spectrum will lack local modes arising during the formation of a Holstein polaron (Melnikov, 1977; Shaw and Whitfield, 1978).

In this case, the values ω_k of the renormalized frequencies of delocalized phonon modes v_k in the case of translation-invariant polarons always lie higher in energy, while for a strong-coupling polaron with spontaneously broken symmetry, they lie below the value ω_k.

The above results indicate that in order to obtain delocalized states that preserve translational invariance, generally speaking, there is no need to arrange spontaneous symmetry breaking in the initial unquantized Hamiltonian of the system, that is, there is no need for the procedure suggested by Higgs (1964) for the introduction of the mass of elementary particles. It is also shown that the study of the extrema of the corresponding classical Hamiltonian cannot provide information as to in the vicinity of which extremum the quantum problem should be solved. In the example of the Holstein Hamiltonian considered by us, such an extremum is the maximum of the classical Hamiltonian. The situation is similar in the case of the Pekar–Fröhlich polaron (Lakhno, 2010b; Lakhno, 2012a, 2013; Kashirina et al., 2012).

3.7 Translation-invariant bipolaron in the Holstein model

The question of the possibility of superconductivity in quasi-one-dimensional systems – polymers and biopolymers – has long attracted the attention of researchers (Williams et al., 1992; Ishiguro et al., 1998; Toyota et al., 2007; Lebed, 2008; Altmore and Chang, 2013). At present, the bipolaron mechanism is considered as one of the possible mechanisms of superconductivity. It is believed that in three-dimensional systems, bipolaron gas forms a Bose condensate with superconducting properties. It is well known that the conditions for the formation of bipolarons in one- and two-dimensional systems are more favorable than in the three-dimensional case. The main problem in this case is the fact that Bose condensation is impossible in one- and two-dimensional systems (Ginzburg, 1968).

In the previous chapter, the concept of translation-invariant polarons and bipolarons was introduced. Under certain conditions, such quasiparticles can possess superconducting properties even in the absence of their Bose condensation. Chapter 2 discussed three-dimensional translation-invariant polarons and bipolarons. In connection with the above, it is of interest to consider the conditions for the formation of translation-invariant bipolarons in low-dimensional systems. In this section, the results obtained above are applied to the quasi-one-dimensional case corresponding to the Holstein model of the polaron.

In the previous section, a translation-invariant theory of 1D polaron was constructed based on the Holstein Hamiltonian. In the case of two electrons, the Holstein Hamiltonian in the one-dimensional case in the continuum limit has the form:

$$H = -\frac{1}{2m}\Delta_{x_1} - \frac{1}{2m}\Delta_{x_2} + \sum_k \left[V_k\left(e^{ikx_1} + e^{ikx_2}\right) a_k + \text{H.c.}\right] + \sum_k \hbar\omega_k^0 a_k^+ a_k + U(x_1 - x_2),$$

$$V_k = \frac{g}{\sqrt{N}}, \quad \omega_k^0 = \omega_0, \tag{3.7.1}$$

where $U(x_1 - x_2)$ is the energy of Coulomb repulsion of two electrons such that:

$$U(x_1 - x_2) = \Gamma\delta(x_1 - x_2), \tag{3.7.2}$$

where Γ is a certain constant and $\delta(x)$ is a delta function.

In the case of broken translational invariance, the bipolaron state is described by a localized wave function $\Psi = \Psi(x_1, x_2)$ and in the strong coupling limit, the total energy functional $\bar{H} = \langle\Psi|H|\Psi\rangle$ according to Kashirina and Lakhno (2010) is determined by the expression:

$$\bar{H} = -\frac{1}{2m}\sum_{i=1,2}\langle\psi|\Delta_{x_i}|\psi\rangle - \sum \frac{V_k^2}{\hbar\omega}|<\psi|e^{ikx_1} + e^{ikx_2}|\psi>|^2 + \langle\psi|U(x_1 - x_2)|\psi\rangle. \tag{3.7.3}$$

The exact solution to problem (3.7.3) is a complex computational problem (Kashirina and Lakhno, 2014; Emin et al., 1992). For the purposes of this section, however, to illustrate the properties of the ground state with broken symmetry, we will use the direct variational method. For this purpose, we choose a trial wave function in the form $\Psi(x_1, x_2) = \varphi(x_1)\varphi(x_2)$. Note that the choice of the probe function in this form corresponds to the exact solution of problem (3.7.3) in the case $U = 0$, that is, in the absence of Coulomb interaction between electrons.

As a result, for the functional of the ground-state energy from (3.7.3) we obtain:

$$\bar{H} = \frac{1}{m}\int |\nabla_x\varphi(x)|^2 dx - \left(\frac{4g^2 a_0}{\hbar\omega_0} - \Gamma\right)\int |\varphi(x)|^4 dx, \tag{3.7.4}$$

where a_0 is a lattice constant. Variation of (3.7.4) with respect to $\varphi(x)$ under normalizing condition leads to the Schrödinger equation:

$$\frac{\hbar^2}{m}\Delta_x\varphi + 2\left(\frac{4g^2a_0}{\hbar\omega_0} - \Gamma\right)|\varphi^2|\varphi + W\varphi = 0, \tag{3.7.5}$$

whose solutions have the form:

$$\varphi(x) = \pm\left(\sqrt{2}rch\frac{x - x_0}{r}\right)^{-1}, \quad r = \frac{2\hbar^2}{m}\frac{1}{((4g^2a_0)/(\hbar\omega_0) - \Gamma)},$$
$$W = -\frac{1}{2}\left(\frac{4g^2a_0}{\hbar\omega_0} - \Gamma\right)^2\frac{m}{2\hbar^2}, \quad E_{bp} = -\frac{1}{6}\left(\frac{4g^2a_0}{\hbar\omega_0} - \Gamma\right)^2\frac{m}{2\hbar^2}, \tag{3.7.6}$$

where x_0 is an arbitrary constant and $E_{bp} = \bar{H}$ is the bipolaron ground-state energy. It should be noted that the polaron ground-state energy E_p is given by expression (3.5.9).

Let us introduce the notation:

$$\gamma = \Gamma\hbar\omega_0/a_0g^2. \tag{3.7.7}$$

It follows from (3.7.6) that for

$$\gamma > 4 \tag{3.7.8}$$

the existence of a bipolaron is impossible. In the case:

$$2 < \gamma < 4 \tag{3.7.9}$$

the metastable bipolaron state decays into separate polaron states. For

$$\gamma < 2 \tag{3.7.10}$$

the bipolaron state will be stable.

Note that the choice of more complex probe functions will not change the qualitative picture, changing only the numerical coefficients (3.7.8)–(3.7.10).

In view of the arbitrary position of the center of mass of a bipolaron x_0 the discussed state of the bipolaron has infinite degeneracy. Any arbitrarily small perturbation of the lattice leads to the elimination of the degeneracy and localization of the bipolaron state at a defect with an attractive potential. A qualitatively different situation is realized in the case of a translation-invariant bipolaron, considered in the next section.

3.8 Translation-invariant bipolaron theory

In the system of the center of mass Hamiltonian (3.7.1) takes the form:

$$H = -\frac{\hbar^2}{2M_e}\Delta_R - \frac{\hbar^2}{2\mu_e}\Delta_r + \sum_k 2V_k \cos\frac{kr}{2}\left(e^{ikR} + \text{H.c.}\right) + \sum_k \hbar\omega_k^0 a_k^+ a_k + U(r)$$

$$R = (x_1 + x_2)/2, \quad r = (x_1 - x_2), \quad M_e = 2m, \quad \mu_e = m/2. \tag{3.8.1}$$

In what follows, we will use units, putting $\hbar = 1$, $\omega_0 = 1$, $M_e = 1$ (accordingly $\mu_e = 1/4$).

The coordinate of the center of mass R in Hamiltonian (3.8.1) can be eliminated via Heisenberg canonical transformation:

$$\hat{S}_1 = \exp\left\{-i\sum_k k a_k^+ a_k R\right\}, \tag{3.8.2}$$

As a result, the transformed Hamiltonian: $\tilde{H} = \hat{S}_1^{-1} H \hat{S}_1$ is written as follows:

$$\tilde{H} = -2\Delta_r + \sum_k 2V_k \cos\frac{kr}{2}\left(a_k^+ + a_k\right) + \sum_k a_k^+ a_k + U(r) + \frac{1}{2}\left(\sum_k k a_k^+ a_k\right)^2. \tag{3.8.3}$$

From (3.8.3) it follows that the exact solution of the bipolaron problem is determined by the wave function $\psi(r)$ which depends only on the relative coordinates r and, therefore, is automatically translation invariant.

Averaging Hamiltonian (3.8.3) over $\psi(r)$ we will write the averaged Hamiltonian as follows:

$$\bar{\tilde{H}} = \bar{T} + \sum_k \bar{V}_k\left(a_k^+ + a_k\right) + \sum_k a_k^+ a_k + \frac{1}{2}\left(\sum_k k a_k^+ a_k\right)^2 + \bar{U},$$

$$\bar{V}_k = 2V_k\langle\psi|\cos\frac{kr}{2}|\psi\rangle, \quad \bar{U} = \langle\psi|U(r)|\psi\rangle, \quad \bar{T} = -2\langle\psi|\Delta_r|\psi\rangle \tag{3.8.4}$$

Subjecting Hamiltonian (3.8.4) to Lee–Low–Pines transformation:

$$\hat{S}_2 = \exp\left\{\sum_k f_k\left(a_k - a_k^+\right)\right\}, \tag{3.8.5}$$

we get:

$$\tilde{\bar{\tilde{H}}} = \hat{S}_2^{-1}\bar{\tilde{H}}\hat{S}_2, \quad \tilde{\bar{\tilde{H}}} = H_0 + H_1, \tag{3.8.6}$$

where

$$H_0 = \bar{T} + 2\sum_k \bar{V}_k f_k + \sum_k f_k^2 + \frac{1}{2}\left(\sum_k k f_k\right)^2 + \bar{U} + H_{KB}, \tag{3.8.7}$$

$$H_{KB} = \sum_k \omega_k a_k^+ a_k + \frac{1}{2}\sum_{k,k'} kk' f_k f_{k'} \left(a_k a_{k'} + a_k^+ a_{k'}^+ + a_k^+ a_{k'} + a_k a_{k'}^+\right), \tag{3.8.8}$$

$$H_1 = \sum_k \left(V_k + f_k \omega_k\right)\left(a_k + a_k^+\right) + \sum_{k,k'} kk' f_{k'} \left(a_k^+ a_k a_{k'} + a_k^+ a_{k'}^+ a_k\right)$$
$$+ \frac{1}{2}\sum_{k,k'} kk' a_k^+ a_{k'}^+ a_k a_{k'}, \tag{3.8.9}$$

$$\omega_k = \omega_0 + \frac{k^2}{2} + k\sum_{k'} k' f_{k'}^2. \tag{3.8.10}$$

According to Appendix I, contribution of $\hat{H}_1$ into the energy vanishes if the eigenfunction of Hamiltonian H_0 transforming the quadratic form H_{KB} to the diagonal one, is chosen properly. Diagonalization of H_{KB} leads to the total energy of the addition ΔE:

$$\Delta E = \frac{1}{2}\sum_k (\nu_k - \omega_k) = -\frac{1}{8\pi i}\int_C \frac{ds}{\sqrt{s}} \ln D(s), \tag{3.8.11}$$

where ν_k are phonon frequencies renormalized by the interaction with the electron. In the one-dimensional case under consideration:

$$D(s) = 1 - \frac{1}{\pi}\int_{-\infty}^{\infty} \frac{k^2 f_k \omega_k}{s - \omega_k^2} dk. \tag{3.8.12}$$

The contour of integration C involved in (3.8.11) is the same as in Chapter 2.

Repeating calculations carried out in Section 3.3 in the one-dimensional case, we express ΔE as follows:

$$\Delta E = \frac{1}{4\pi}\int_{-\infty}^{\infty} \frac{k^2 f_k^2 dk}{2(1+Q)} + \frac{1}{4\pi^2}\int\int_{-\infty}^{\infty} \frac{k^2 f_k^2 p^2 f_p^2 \omega_p \left(\omega_k \omega_p + \omega_k\left(\omega_k + \omega_p\right) + 1\right)}{\left(\omega_k + \omega_p\right)^2 \left(\omega_p^2 - 1\right) \left|D_+\left(\omega_p^2\right)\right|^2} dp dk,$$

$$D_+\left(\omega_p^2\right) = 1 + \frac{1}{\pi}\int_{-\infty}^{\infty} \frac{k^2 f_k^2 \omega_k dk}{\omega_k^2 - \omega_p^2 - i\varepsilon}. \tag{3.8.13}$$

Finally, with the use of (3.8.7) and (3.8.8), the bipolaron total energy E is written as follows:

$$E = \Delta E + 2\sum_k \bar{V}_k f_k + \sum_k f_k^2 + \bar{T} + \bar{U}. \tag{3.8.14}$$

3.9 Variational calculation of the bipolaron state

We could have derived exact equations for determining the bipolaron energy by varying (3.8.14) with respect to ψ and f_k. The quantities ψ and f_k obtained as solutions of this equation, being substituted into (3.8.14) determine the bipolaron total energy E. Since finding a solution of the equation obtained by variation of E is rather a complicated procedure, we will use the variational approach. To this end, let us choose the probe functions ψ and f_k in the form:

$$\psi(r) = \left(\frac{2}{\pi}\right)^{1/4} \frac{1}{\sqrt{l}} e^{-r^2/l^2}, \tag{3.9.1}$$

$$f_k = -Nge^{-k^2/2a^2}, \tag{3.9.2}$$

where N, l, and a are variational parameters. Substitution of (3.9.1) and (3.9.2) into (3.8.14) after minimization with respect to N leads to the following expression for E:

$$E_{bp} = \frac{ma_0^2}{\hbar^2} \frac{g^4}{\hbar^2 \omega_0^2} min(x,y)E(x,y;\gamma), \tag{3.9.3}$$

$$E(x,y;\gamma) \approx 2\left(0.390625x^2 + \frac{2}{y^2} - \frac{4x}{\sqrt{\pi}(1+x^2y^2/16)} + \sqrt{\frac{2\gamma}{\pi y}}\right). \tag{3.9.4}$$

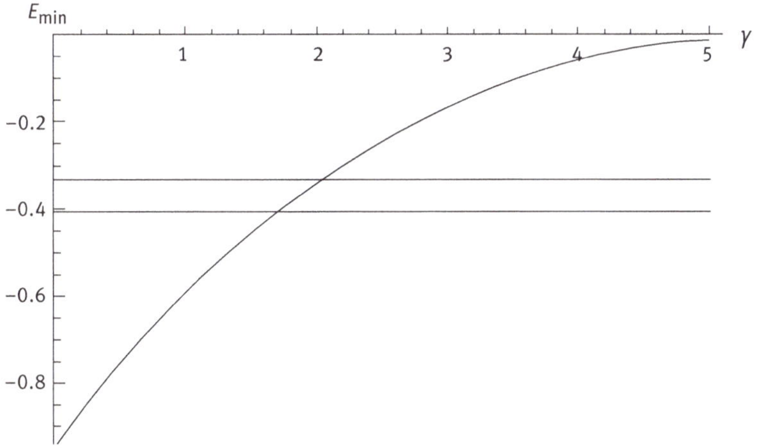

Fig. 3.2: Dependence of E_{min} on γ.

The expression for the bipolaron energy is given in dimension units. The results of minimization of function $E(x,y,\gamma)$ with respect to dimensionless parameters x, y are presented in Fig. 3.2 for various values of the parameter γ. Figure 3.2 suggests that as

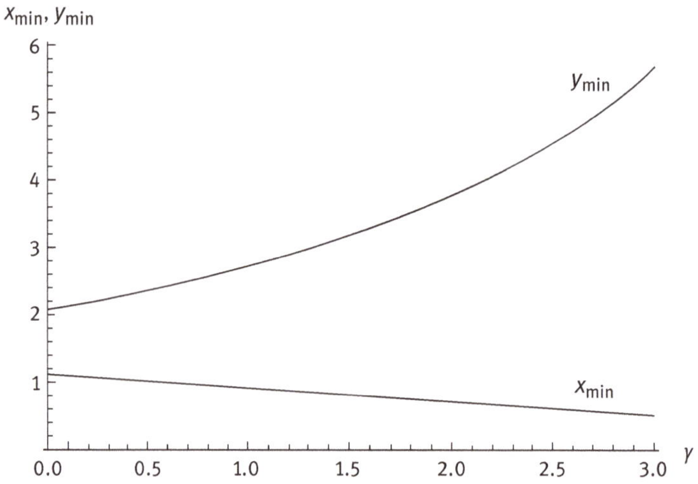

Fig. 3.3: Dependence of x_{min}, y_{min} on γ.

distinct from a bipolaron with broken symmetry (inequality (3.7.8)), a translation-invariant bipolaron exists for all the values of the parameter γ.

In the region:

$$\gamma > 3.02 \tag{3.9.5}$$

a translation-invariant bipolaron is stable relative to its decay into two individual Holstein polarons, but remains unstable relative to decomposition into individual translation-invariant polarons.

For

$$\gamma < \gamma_c = 2.775 \tag{3.9.6}$$

a translation-invariant bipolaron becomes stable relative to its decay into two individual polarons. Notice that for $\gamma = 0$, the energy of a translation-invariant bipolaron is equal to: $E_{bp} = -1.87104ma_0^2g^4/\hbar^4\omega_0^2$, that is, lies much lower than the exact value of the energy of a bipolaron with broken symmetry, which, according to (3.7.6), is equal to $E_{bp} = -(4/3)ma_0^2g^4/\hbar^4\omega_0^2$. The energy of a translation-invariant bipolaron also lies below the variational estimate of the energy of a bipolaron with spontaneously broken symmetry (3.7.6) for all the values of γ (Kashirina and Lakhno, 2014).

The dimensionless parameters x, y involved in (3.9.4) are related to the variational parameters a, l (3.9.1) and (3.9.2) as follows: $a = \left(2ma_0^2g^2/\hbar^3\omega_0\right)x$ and $l = \left(\hbar^3\omega_0/2ma_0^2g^2\right)y$. The parameter l determines the characteristic size of the electron pair, that is, the correlation length $L(\gamma)$, whose dependence on γ is given by the expression:

$$L(\gamma) = \frac{\hbar^2}{2ma_0^2}\frac{\hbar\omega_0}{g^2}y_{min}(\gamma) \tag{3.9.7}$$

The dependencies of y_{min} and x_{min} on y are presented in Fig. 3.3. Figure 3.3 suggests that the correlation length $L(\gamma)$ in the region of a bipolaron stability $0 < \gamma < \gamma_c$ does not change greatly and for its critical value $\gamma_c = 2.775$ the quantity $L(\gamma)$ approximately three times exceeds the value of $L(0)$, that is, the correlation length in the absence of the Coulomb repulsion. This qualitatively differs from the case of a bipolaron with broken symmetry for which the corresponding value, according to (3.7.6), for $\gamma = \gamma_c$ turns to infinity.

3.10 On the possibility of bipolaron states in discrete chains

To describe electrons in a discrete one-dimensional molecular chain, Hubbard model (Hubbard, 1963) is used, whose Hamiltonian has the form (Holstein, 1959a, 1959b; Proville and Aubry, 1998; Lakhno, 2006):

$$\hat{H} = \eta \sum_{(i,j),\sigma} c_{i\sigma}^{+} c_{j\sigma} + \sum_j \hbar\omega\left(a_j^{+} a_j + \frac{1}{2}\right) + \sum_j g\hat{n}_j\left(a_j^{+} + a_j\right) + \sum_j U\hat{n}_{j\uparrow}\hat{n}_{j\downarrow}, \tag{3.10.1}$$

where $\hat{n}_{j\sigma} = c_{j\sigma}^{+}c_{j\sigma}$, $\hat{n}_j = \hat{n}_{j\uparrow} + \hat{n}_{j\downarrow}$, $c_{j\sigma}^{+}$, $c_{j\sigma}$ are operators of the birth and annihilation of an electron with a spin σ on the jth site; η is a matrix element of the transition between neighboring sites (i, j); g is a constant of an electron interaction with the chain oscillations; U is a parameter of Coulomb repulsion; and $\hbar\omega$ is the energy of oscillator vibrations. By a site here we mean a diatomic molecule considered as an oscillator.

In the one-dimensional case, regardless of the specific type of molecular chain, bipolaron states for Hamiltonian (3.10.1) were considered in Proville and Aubry (1998). However, the stability of bipolaron states with respect to their decay into individual polaron states was not analyzed in it (Lakhno and Sultanov, 2011).

According to Proville and Aubry (1998), the ground state of Hamiltonian (3.10.1) is the wave function of the form:

$$|\psi\rangle = \sum_{ij} \psi_{ij} c_{i\uparrow}^{+} c_{j\downarrow}^{+} |0\rangle, \tag{3.10.2}$$

where $|0\rangle$ is the vacuum wave function, $\psi_{ij} = \psi_{ji}$, $\sum_{ij} |\psi_{ij}|^2 = 1$. In the adiabatic approximation, when $1/4(\hbar\omega/2g)^4 \ll 1$, wave functions ψ_{ij} satisfy the equation:

$$\eta(\Delta\psi)_{ij} + \left(-\kappa\left(\rho_i + \rho_j\right) + U\delta_{i-j}\right)\psi_{ij} = W\psi_{ij}, \tag{3.10.3}$$

$$\rho_i = 1/2\sum_k \left(|\psi_{ik}|^2 + |\psi_{ki}|^2\right),$$

where $\kappa = 4g^2/\hbar\omega$, Δ is the discrete Laplace operator:

$$(\Delta\psi)_{ij} = \psi_{i-1,j} + \psi_{i+1,j} + \psi_{i,j-1} + \psi_{i,j+1}.$$

Notice that for $\kappa = 0$, $U = 0$, $N \to \infty$, where N is the number of sites in a chain, the spectrum of permissible energies W is a closed interval $[-4\eta, 4\eta]$, where -4η is the doubled value of the bottom of the conductivity band of a hole in the chain.

Localized hole states correspond to energies W, which lie below the doubled energy corresponding to the bottom of the conductivity band.

It follows from (3.10.1) to (3.10.3) that the total energy of a bipolaron state reckoned from bottom of the conductivity band is equal to:

$$E_{bp} = W + \frac{\kappa}{2}\sum_{i,j} |\psi_{ij}|^2 \left(\rho_i + \rho_j\right) + 4\eta. \tag{3.10.4}$$

The condition for the stability of bipolaron states is determined by the inequality:

$$|E_{bp}| > 2|E_p|, \tag{3.10.5}$$

where

$$E_p = \lambda + \frac{\kappa}{4}\sum_i |\psi_i|^4 + 2\eta \tag{3.10.6}$$

and $\psi_i\left(\sum_i |\psi_i|^2 = 1\right)$ has the meaning of the wave function of a one-polaron ground state of an electron with relevant energy λ, determined by the equation:

$$\eta(\Delta\psi)_i - \frac{\kappa}{2}|\psi_i|^2\psi_i = \lambda\psi_i, \qquad (\Delta\psi)_i = \psi_{i-1} + \psi_{i+1}. \tag{3.10.7}$$

As an example, let us consider a polynucleotide chain in which the role of a site is played by the complementary Watson–Crick pair G/C, where G is guanine and C is cytosine. The charge carriers in such a chain are holes (Offenhäusser and Rinaldi, 2009).

For the poly(G)/poly(C) duplex, the parameter values $\eta = 0.084$ eV and $\kappa = 0.5267$ eV are taken the same as in Lakhno (2010a). In the considered case of charged particles (holes), the parameter $U \neq 0$. Its exact value is unknown, but an approximate value can be obtained from the following considerations. For two holes localized on one nucleotide, the energy of the Coulomb repulsion is approximately equal to $e^2/2\varepsilon_\infty l$, where ε_∞ is the high-frequency dielectric constant of the molecule; l is the characteristic size of the nucleotide. For DNA, these quantities are of the order of magnitude $\varepsilon_\infty = 2$ (Balabaev and Lakhno, 1991; Lakhno, 2006), $l = 3$ Å, which gives the value of $U \approx 1$ eV. With this value of the U parameter, localization on one site is not possible. The state when holes are localized at two adjacent sites or at a greater number of sites has a lower energy value.

The ground bipolaron and polaron states, that is, solutions of equations (3.10.3) and (3.10.7) were calculated by minimizing functionals $E_{bp}\left(\psi_{ij}\right)$ (3.10.4) and $E_p(\psi_i)$ (3.10.6) with symmetry and normalization conditions for η= 0.084 eV in the range of parameters κ, U, containing the above values for the duplex poly(G)/poly(C). Figure 3.4 shows the diagram of the stability of bipolaron states, determined by inequality (3.10.5), obtained by interpolation by the least squares method over 40 calculated points.

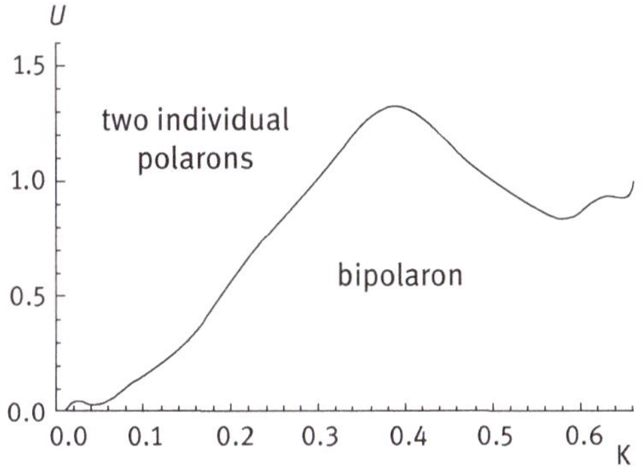

Fig. 3.4: Region of stability $|E_{bp}| > 2|E_p|$ of a bipolaron; $\eta = 0.084$ eV.

From (3.10.3), (3.10.4) and (3.10.6), (3.10.7) for $\kappa \to \infty$ we have:

$$|E_{bp}| = |U - \kappa| = U - \kappa > 2|E_p| = 2| - \kappa/4| = \kappa/2.$$

Hence, at large values of κ the border of the region is asymptotically serves the straight line: $U = (3/2)\kappa$, which can be seen from the numerical calculation in Fig. 3.4. For $\kappa \to \infty$ obviously $U \to 0$, $U < \kappa$.

It follows from (3.10.3) and (3.10.4) that for $U = \infty$, that is when the holes are at a great distance from each other, the bipolaron state decays into two individual polarons. In this case, in the continuum limit, the total energy of the bipolaron (3.10.4) is eight times lower than the corresponding energy of the polaron (3.10.6): $E_{bp}/E_p = 8$. By virtue of the virial theorem, this relation is also fulfilled for the eigenenergies of the polaron and bipolaron: $W/\lambda = 8$. In the discrete case, one can only assert that if we take a double value of κ in the polaron equation (3.10.7), then we will obtain an eigenvalue that is twice as small in absolute value as that of a bipolaron.

Figure 3.5a–d shows charge densities $p_i = 2\sum_j \left|\psi_{ij}\right|^2$ of different types of bipolarons for $\eta = 0.084$ eV, $\kappa = 0.5267$ eV, $N = 31$. The solution shown in Fig. 3.5 with $U = 0$ could correspond to Frenkel biexciton self-localized states. Notice that in the case of

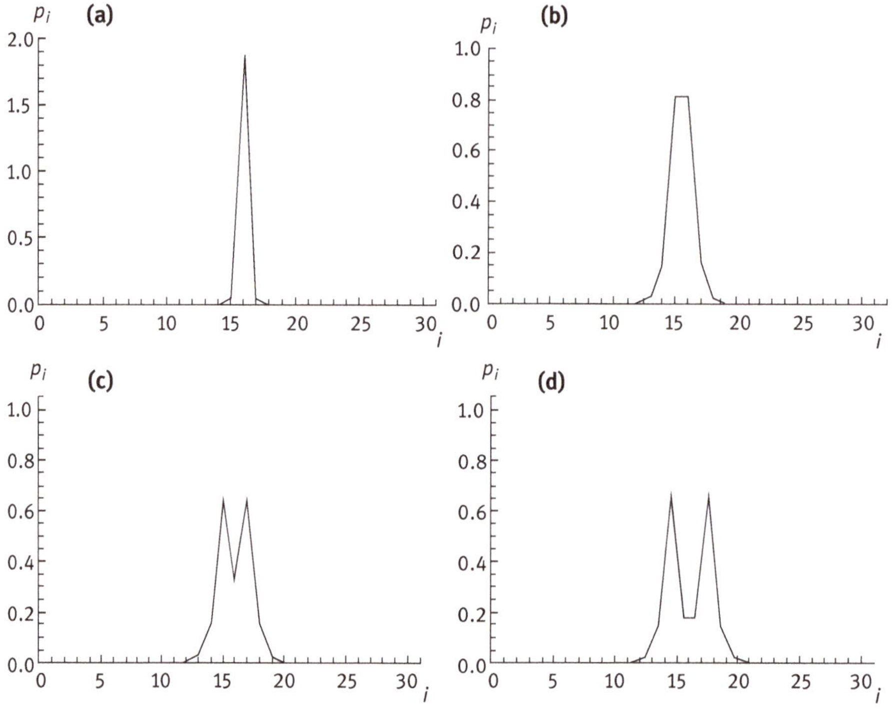

Fig. 3.5: Charge densities (p_i) of different-type bipolarons for (a) $U = 0$; (b) $U = 0.33\text{eV}$; (c) $U = 0.6\text{eV}$; (d) $U = 0.75\text{eV}$, where U is the Coulomb repulsion parameter.

$U = 0$, the bipolaron state is stable for any parameter values, and the lowest energy state corresponds to a two-particle state localized at one site.

Expressions (3.10.3)–(3.10.7) suggest that for $\kappa \geq 4U$ the ground state corresponds to localization of the holes at one site. For $\kappa \leq 4U$ their localization at two or more sites becomes energetically advantageous (Fig. 3.5c, d). In the latter case for $\kappa > 4\eta$ the state is localized at two sites. In the range of values of the parameters $\kappa > 4\eta$, $\kappa > 4U$ the holes are localized at different sites. The energy of the holes localized at two neighboring sites differs only slightly from the energy of the holes localized at two sites far apart from each other. It can be said that the bipolaron state in this case is weakly stable relative to decay into two individual polaron states. In the range of values $\kappa \leq 4\eta$ the bipolaron state is localized at more than two sites.

Figure 3.5a–d suggests that in the region of stability the bound bipolaron state is formed by polaron occurring at distances of 0, 1, 2, and 3 lattice constants.

To quantitatively estimate the length of bipolaron states, we introduce the quantities P_1 and P_2:

$$P_1 = \sum_i |i - i_0| p_{i-i_0}, \quad P_2 = \frac{1}{\sum_i p_i^2},$$

i_0 is the center of symmetry of a bipolaron and p_i is a charge density.

P_1 determines a characteristic size of a bipolaron, P_2 determines a characteristic size of a polaron (characteristic size of peaks in a two-center configuration). For $U = 0$, when a one-center configuration takes place, P_1 and P_2 are of the same order of magnitude. For bipolarons in Fig. 3.5, the values of these quantities are equal: $P_1 = 1.65$, $P_2 = 0.28$ (Fig. 3.5a); $P_1 = 17.5$, $P_2 = 0.72$ (Fig. 3.5b); $P_1 = 24.7$, $P_2 = 1$ (Fig. 3.5c); $P_1 = 26.1$, $P_2 = 1.04$ (Fig. 3.5d).

The table lists the bipolaron binding energies $\Delta = |E_{bp}| - 2|E_p|$ for some values of the parameters from the stability region Fig. 3.4.

Table 3.1 suggests that when a bipolaron is maximally stable, that is, when $U = 0$, the bipolaron coupling energy for $\kappa = 0.5267$ eV is $\Delta = |E_{bp}| - 2|E_p| = 0.203$ eV. As U increases, the bipolaron binding energy decreases.

Tab. 3.1: Bipolaron binding energies $\Delta = |E_{bp}| - 2|E_p|$ ($\eta = 0.084$ eV).

$U = 1$	*	*	0.000025	0.000125	*	*
$U = 0.75$	*	*	0.0001	0.00045	0.00047	0.001
$U = 0.5$	*	*	0.0005	0.0018	0.0027	0.0097
$U = 0.25$	*	0.00047	0.0028	0.009	0.022	0.062
$U = 0$	0.0037	0.015	0.05	0.112	0.203	0.283
k	0.1	0.1975	0.296	0.395	0.5267	0.66

Note. *Bipolaron binding energies Δ in eV (* – bipolaron is unstable: $\Delta < 0$). Values of U and κ in eV.*

The results obtained testify to the possibility of bipolaron states in homogeneous nucleotide chains.

As it follows from Fig. 3.4, near the value $\kappa = 0.5267$ eV, there is a big local minimum on the phase diagram where the value of U although is less than 1 lies close to this value. Since the value of U was assessed only approximately, it is possible that the exact value of this quantity will satisfy the condition for the formation of bipolaron states. There is currently no exact value for κ either. Moreover, its value can vary greatly depending on external conditions.

The value $\kappa \approx 0.5$ eV corresponds to a “dry” DNA molecule. It is seen from Tab. 3.1 that for $\kappa = 0.4 \div 0.5$ eV and $U = 0.75$ eV the value $\Delta \approx 5 \times 10^{-4}$ eV, which corresponds to the temperature of a superconducting transition $T_C \approx 6$ K. In a solution, the value of κ can greatly increase (according to Voityuk (2005a, 2005b) approximately five times). In this case, with a very high probability, the conditions for the formation of bipolarons will be fulfilled. In any case, the results obtained demonstrate that both on-site and inter-site bipolarons have very small bounding energy and for this reason can explain only low-temperature superconductivity.

Note that earlier, attention was drawn to the possibility of bipolaron states in short (six-site) heterogeneous oligonucleotide chains in Apalkov and Chakraborty (2006). According to Apalkov and Chakraborty (2006), the localization of pairs in such chains should occur on doublets, triplets, and quadruplets of guanines. The main problem associated with the possibility of a superconducting state in DNA is the need for the existence of current carriers (electrons or holes Lakhno ,2008; Lewis and Wu, 2001; Okahata et al.,1998; Porath et al., 2000) in the molecule, since DNA itself is a dielectric (de Pablo et al., 2000; Storm et al., 2001; Yoo et al., 2001). In the experiment (Kasumov et al., 2001), when measuring the conductivity, a DNA molecule was attached to two rhenium-carbon electrodes lying on a mica substrate. The gap between the electrodes in Kasumov et al. (2001) was created by burning out a strip ($\approx 0.5 \mu m$ thick) by a laser, which was mica with islands of rhenium–carbon atoms remaining unconnected with each other on it. In our opinion, the appearance of current carriers in DNA could be due to the contact of the molecule attached to the substrate with the rhenium–carbon islands of the gap.

4 Translational-invariant bipolarons and superconductivity

A translation-invariant (TI) bipolaron theory of superconductivity based, like Bardeen–Cooper–Schrieffer (BCS) theory, on Fröhlich Hamiltonian is presented. Here the role of Cooper pairs belongs to TI bipolarons which are pairs of spatially delocalized electrons whose correlation length of a coupled state is small. The presence of Fermi surface leads to the stabilization of such states in its vicinity and a possibility of their Bose–Einstein condensation.

The theory provides a natural explanation of the existence of a pseudogap phase preceding the superconductivity and enables one to estimate the temperature of a transition T^* from a normal state to a pseudogap one. It is shown that the temperature of BEC of TI bipolarons determines the temperature of a superconducting transition T_c which depends not on the bipolaron effective mass but on the ordinary mass of a band electron. This removes restrictions on the upper limit of T_c for a strong electron–phonon interaction (EPI). A natural explanation is provided for the angular dependence of the superconducting gap which is determined by the angular dependence of the phonon spectrum.

It is demonstrated that a lot of experiments on thermodynamic and transport characteristics, Josephson tunneling and angle-resolved photoemission spectroscopy of high-temperature superconductors does not contradict the concept of a TI bipolaron mechanism of superconductivity in these materials. Possible ways of enhancing T_c and producing new room-temperature superconductors are discussed on the basis of the theory suggested.

4.1 Weak EPI: BCS theory

In BCS a multielectron problem is solved on the assumption that electrons interact only with a phonon field and do not interact with one another. Hence, only an ensemble of independent electrons in a phonon field is considered. Such a picture of BCS is substantiated by a Fermi-liquid model of a metal which implies that instead of strongly interacting electrons we can consider noninteracting quasiparticles, that is, an ideal Fermi gas in a phonon field. In this case, the one-electron Fröhlich Hamiltonian (1.2.1) can be written in the form suitable for the description of any number of electrons:

$$H = \sum_{p,s} \varepsilon_p c^+_{p,s} c_{p,s} + \sum_q \omega(q) a^+_q a_q + \sum_{\substack{p,q,s \\ p'-p=q}} V_q c^+_{p,s} c_{p',s} a^+_q + H.c. \tag{4.1.1}$$

$$\varepsilon_p = p^2/2m - E_F, \qquad \omega(q) = s_0 q,$$

where $c^+_{p,s}$, $c_{p,s}$ are operators of the birth and annihilation of electrons with momentum p and spin s, s_0 is the sound velocity. In (4.1.1), the energy of electron states is reckoned from the Fermi level E_F.

https://doi.org/10.1515/9783110786668-004

In the case of metals for which the BCS is used:

$$V_q = G(\omega(q)/2V)^{1/2},$$

G is the interaction constant. For a weak EPI, using the perturbation theory we can exclude phonon operators a_q^+, a_q and present (4.1.1) as the Hamiltonian:

$$H = \sum_{p,s} \varepsilon_p c_{p,s}^+ c_{p,s} + \sum_{p,p',k,s,s'} V_k^2 \frac{\hbar\omega(k)}{(\varepsilon_{p+k} - \varepsilon_k)^2 - \hbar^2\omega^2(k)} c_{p+k,s}^+ c_{p'-k,s}^+ c_{p',s'} c_{p,s} \tag{4.1.2}$$

In the BCS theory, an important approximation is made: it is believed that the main contribution into the interaction is made only by the processes occurring in the energy range $|\varepsilon_p - \varepsilon_{p'}| < \hbar\omega_D$ in the vicinity of the Fermi level where ω_D is the Debye frequency of a phonon. In this energy range the coefficient preceding the electron operators in the interaction term is replaced by the constant g.

The BCS theory is based on the choice of a probe function in the form of a superposition of Cooper pairs with $p = -p'$, $s = -s'$. Hence, in the BCS instead of (4.1.2) consideration is given to the Hamiltonian:

$$\begin{aligned} H &= \sum_{p,s} \varepsilon_p c_{p,s}^+ c_{p,s} - g \sum_{p,k} c_{p+k,\uparrow}^+ c_{-p-k,\downarrow}^+ c_{-p,\downarrow} c_{p,\uparrow} \\ &= \sum_{p,s} \varepsilon_p c_{p,s}^+ - g \sum_{p} c_{p,\uparrow}^+ c_{-p,\downarrow}^+ \sum_{p'} c_{-p',\downarrow} c_{p',\uparrow} \end{aligned} \tag{4.1.3}$$

Hamiltonian (4.1.3) can be diagonalized via the canonical transformation:

$$\begin{aligned} c_{p,\uparrow} &= u_p \xi_{p,\uparrow} + v_p \xi_{-p,\downarrow}^+, \quad c_{-p,\downarrow} = u_p \xi_{-p,\downarrow} + v_p \xi_{p,\uparrow}^+, \\ c_{p,\uparrow}^+ &= u_p \xi_{p,\uparrow}^+ + v_p \xi_{-p,\downarrow}, \quad c_{-p,\downarrow}^+ = u_p \xi_{-p,\downarrow}^+ - v_p \xi_{p,\uparrow}. \end{aligned} \tag{4.1.4}$$

As a result, Hamiltonian (4.1.3) is written as

$$H = E_0 + \sum_p E_p \left(\xi_{p,\uparrow}^+ \xi_{p,\uparrow} + \xi_{-p,\downarrow}^+ \xi_{-p,\downarrow} \right) \tag{4.1.5}$$

$$E_p = \sqrt{\left(\frac{p^2}{2m} - E_F\right)^2 + \Delta^2}, \quad \Delta = 2g\sum_p{}' u_p v_p, \quad u_p^2 = 1 - v_p^2 = \left(1 + \varepsilon_p / E_p\right)/2,$$

where the prime in the expression for Δ means that summation is performed over the states lying in a thin layer of the Fermi surface where interaction is nonzero, $|v_p|^2$ provides a probability that the state $(p,\uparrow, -p,\downarrow)$ is occupied, and $|u_p|^2$ is the probability that it is free.

The results obtained correspond only to the case of $T = 0$. In particular, the energy of the ground state of the system under consideration reckoned from the energy of the system in the normal state (i.e., with $\Delta = 0$) is equal to

$$E = \langle\Psi|H|\Psi\rangle = -\frac{1}{2}N(0)\Delta^2, \quad \Delta = 2\hbar\omega_D \exp\left(-\frac{1}{N(0)g}\right),$$

$$|\Psi\rangle = \prod\nolimits_{p_1\ldots p_N}\left(u_p + v_p c^+_{p,\uparrow} c^+_{-p,\downarrow}\right)|0\rangle, \tag{4.1.6}$$

where $N(0)$ is the electron density at the Fermi level in a normal phase, N is the number of electrons.

Hence, formation of paired states leads to a decrease of the system energy by the value of $N(0)\Delta^2/2$ and emergence of superconductivity. It follows from (4.1.5) that the density of elementary excitations $\rho(E_p) \to \infty$ for $E_p \to \Delta$. In the translation-invariant (TI) bipolaron theory of SC this corresponds to the formation of a Bose condensate of paired electrons with an infinite state density for the energy equal to the bipolaron energy which is separated by a gap from the continuous excitation spectrum.

The problem of the number of paired electrons, that is, Cooper pairs in the BCS theory is treated differently by different authors. For example, it is often argued (see, for example, Weisskopf, 1981), that electrons are paired only in the narrow layer of the Fermi surface so that their number N_s is equal to $N_s = (\Delta/E_F)N$. For $\Delta/E_F \cong 10^{-4}$ only a small portion of electrons are paired.

The BCS theory, however, gives an unambiguous answer: for $T = 0$ $N_s = N/2$ (which straightforwardly follows from the expression for the wave function (4.1.6)), that is, all the electrons are in the paired state.

To resolve this contradiction let us consider the contribution of w_p into the total energy of a superconductor which is made by a pair in the state $(p,\uparrow - p,\downarrow)$:

$$w_p = \varepsilon_p - E_p. \tag{4.1.7}$$

It follows from (4.1.7) that $w_{p_F} = -\Delta$. In the normal state ($\Delta = 0$) such a pair would contribute the energy $w^N_{p_F} = 0$, that is $\delta w_{p_F} = w^s_{p_F} - w^N_{p_F} = -\Delta$. Accordingly, at the bottom of the conductivity band, that is, for $p = 0$ expression (4.1.7) is written as $w^s_0 = -2E_F - \Delta^2/2E_F$. In a normal state such a pair would contribute the energy $w^N_o = -2E_F$, accordingly, $\delta w_0 = w^s_0 - w^N_0 = -\Delta^2/2E_F$. It follows that pairs occurring far below the Fermi surface outside the layer of width Δ, in the BCS approximation make a very small contribution into the SC energy which is approximately $\delta w_0/\delta w_{p_F} = \Delta/2E_F \sim 10^{-4}$ of the contribution made by the pairs in the layer Δ. Hence, though all the electrons in the BCS are paired, their contribution depending on the energy of the pair is different. Only in the thin layer of the Fermi surface it is nonzero. This fact just resolves the above contradiction: though all the electrons are paired, the energy is contributed only by a small number of the pairs: $N_s = N\Delta/E_F$, which is called a number of pairs in a SC.

Therefore, in order not to make a mistake in calculating some or other characteristic of a SC, when a solution is not obvious, one should take account of all the paired states of the electrons.

For example, when calculating London depth of the magnetic field penetration into a SC in the BCS theory, one should take account of all the electron paired states. At the same time, when calculating the critical magnetic field for which an SC deteriorates, it is sufficient to estimate them only in the layer Δ.

It should be noted that in recent experiments by Božović et al. (2016) it was shown that only a small portion of electrons make a contribution into the London penetration depth in HTSC. This means that the BCS theory is inapplicable to them and the interaction cannot be considered to be weak. This problem will be considered in Section 4.9.

4.2 Pekar–Fröhlich Hamiltonian: canonical transformations

Before we pass on to presentation of the SC theory in the limit of strong EPI, let us outline the results of the TI bipolaron theory.

In describing bipolarons, according to Lakhno (2010b, 2012a, 2015a, 2019b) and Kashirina et al. (2012), we will proceed from Pekar–Fröhlich Hamiltonian in a magnetic field:

$$H = \frac{1}{2m}\left(\hat{\mathbf{p}}_1 - \frac{e}{c}\mathbf{A}_1\right)^2 + \frac{1}{2m}\left(\hat{\mathbf{p}}_2 - \frac{e}{c}\mathbf{A}_2\right)^2 + \sum_k \hbar\omega_k^0 a_k^+ a_k +$$
$$\sum_k \left(V_k e^{ikr_1} a_k + V_k e^{ikr_2} a_k + H.c.\right) + U(|\mathbf{r}_1 - \mathbf{r}_2|), \tag{4.2.1}$$
$$U(|\mathbf{r}_1 - \mathbf{r}_2|) = \frac{e^2}{\varepsilon_\infty |\mathbf{r}_1 - \mathbf{r}_2|},$$

where $\hat{\mathbf{p}}_1$, $\mathbf{r}_1$, $\hat{\mathbf{p}}_2$, $\mathbf{r}_2$ are momenta and coordinates of the first and second electrons, $\mathbf{A}_1 = \mathbf{A}(\mathbf{r}_1)$, $\mathbf{A}_2 = \mathbf{A}(\mathbf{r}_2)$ are vector-potentials of the external magnetic field at the points where the first and second electrons occur; U describes Coulomb repulsion between the electrons. We write Hamiltonian (4.2.1) in general form. In the case of HTSC which are ionic crystals V_k is a function of the wave vector k, which corresponds to the interactions between the electrons and optical phonons:

$$V_k = \frac{e}{k}\sqrt{\frac{2\pi\hbar\omega_0}{\tilde{\varepsilon}V}} = \frac{\hbar\omega_0}{ku^{1/2}}\left(\frac{4\pi\alpha}{V}\right)^{1/2}, \quad u = \left(\frac{2m\omega_0}{\hbar}\right)^{1/2}, \quad \alpha = \frac{1}{2}\frac{e^2 u}{\hbar\omega_0\tilde{\varepsilon}} \tag{4.2.2}$$
$$\tilde{\varepsilon}^{-1} = \varepsilon_\infty^{-1} - \varepsilon_0^{-1}, \quad \omega_k^0 = \omega_0,$$

where e is the electron charge; ε_∞ and ε_0 are high-frequency and static dielectric permittivities; α is a constant of the EPI; V is the system volume, ω_0 is a frequency of an optical phonon.

The axis z is chosen along the magnetic field induction **B** and symmetrical gauge is used:

$$\mathbf{A}_j = \frac{1}{2}\mathbf{B}\times\mathbf{r}_j,$$

for $j = 1, 2$. For the bipolaron singlet state considered below, the contribution of the spin term is equal to zero.

In the system of the mass center Hamiltonian (4.2.1) takes the form:

$$H = \frac{1}{2M_e}\left(\hat{\mathbf{p}}_R - \frac{2e}{c}\mathbf{A}_R\right)^2 + \frac{1}{2\mu_e}\left(\hat{\mathbf{p}}_r - \frac{e}{2c}\mathbf{A}_r\right)^2 + \sum_k \hbar\omega_k^0 a_k^+ a_k + \sum_k 2V_k \cos\frac{\mathbf{kr}}{2}\left(a_k e^{ikR} + H.c.\right) + U(|\mathbf{r}|)\,, \quad (4.2.3)$$

$$\mathbf{R} = \frac{\mathbf{r}_1 + \mathbf{r}_2}{2}, \quad \mathbf{r} = \mathbf{r}_1 - \mathbf{r}_2, \quad M_e = 2m, \quad \mu_e = m/2, \quad \mathbf{A}_r = 1/2B(-y, x, 0),$$

$$\mathbf{A}_R = 1/2B(-Y, X, 0), \quad \hat{\mathbf{p}}_R = \hat{\mathbf{p}}_1 + \hat{\mathbf{p}}_2 = -i\hbar\nabla_R \quad \hat{\mathbf{p}}_r = (\hat{\mathbf{p}}_1 - \hat{\mathbf{p}}_2)/\mathbf{2} = -i\hbar\nabla_\mathbf{r}$$

where $x; y$ and $X; Y$ are components of the vectors **r**, **R** accordingly.

Let us transform Hamiltonian H by Heisenberg transformation (Heisenberg, 1930; Rosenfeld, 1932):

$$S_1 = \exp\, i\left(\mathbf{G} - \sum_k \mathbf{k} a_k^+ a_k\right)\mathbf{R}, \quad (4.2.4)$$

$$\mathbf{G} = \hat{\mathbf{P}}_R + \frac{2e}{c}\mathbf{A}_R, \; \hat{\mathbf{P}}_R = \hat{\mathbf{p}}_R + \sum_\mathbf{k} \hbar\mathbf{k} a_k^+ a_k, \quad (4.2.5)$$

where **G** commutates with the Hamiltonian, thereby being a constant, that is c-number, is the total momentum in the absence of the magnetic field.

Action of S_1 on the field operator yields:

$$S_1^{-1} a_k S_1 = a_k k e^{-ikR}, \qquad S_1^{-1} a_k^+ S_1 = a_k^+ e^{ikR} \quad (4.2.6)$$

Accordingly, the transformed Hamiltonian $\tilde{H} = S_1^{-1} H S_1$ takes on the form:

$$\tilde{H} = \frac{1}{2M_e}\left(\mathbf{G} - \sum_k \hbar\mathbf{k} a_k^+ a_k - \frac{2e}{c}\mathbf{A}_R\right)^2 + \frac{1}{2\mu_e}\left(\hat{\mathbf{p}}_r - \frac{e}{2c}\mathbf{A}_r\right)^2 + \sum_k \hbar\omega_k^0 a_k^+ a_k + \sum_k 2V_k \cos\frac{\mathbf{kr}}{2}\left(a_k + a_k^+\right) + U(|\mathbf{r}|)\,. \quad (4.2.7)$$

In what follows, we will assume:

$$G = 0. \tag{4.2.8}$$

The physical meaning of condition (4.2.8) is that the total momentum in a sample volume is equal to zero, that is, a current is lacking. This requirement follows from the Meissner effect according to which the current in a SC volume should be equal to zero. We use this fact in Section 4.7 in determining the London penetration depth λ.

Let us seek a solution of a stationary Schrödinger equation corresponding to Hamiltonian (4.2.7) in the form:

$$\Psi_H(r,R,\{a_k\}) = \phi(R)\Psi_{H=0}(r,R,\{a_k\}) \tag{4.2.9}$$

$$\phi(R) = \exp\left(-i\frac{2e}{\hbar c}\int_0^R \mathbf{A}_{R'}d\mathbf{R}'\right), \qquad \Psi_{H=0}(r,R,\{a_k\}) = \Psi(r)\Theta(R,\{a_k\}).$$

where $\Psi_{H=0}(r,R,\{a_k\})$ is the bipolaron wave function in the absence of a magnetic field. The explicit form of the functions $\psi(r)$and $\Theta(R,\{a_k\})$ is given in Lakhno 2010b, 2012a, 2015b) (expression (2.11.13)).

Averaging of $\tilde{H}$ with respect to the wave functions $\phi(R)$ and $\psi(r)$ yields:

$$\bar{\tilde{H}} = \frac{1}{2M_e}\left(\sum_k \hbar\mathbf{k}a_k^+a_k\right)^2 + \sum_k \hbar\tilde{\omega}_k a_k^+ a_k + \sum_k \bar{V}_k(a_k + a_k^+) + \bar{T} + \bar{U} + \bar{\Pi}, \tag{4.2.10}$$

where

$$\bar{T} = \frac{1}{2\mu_e}\langle\psi|\left(\hat{\mathbf{p}}_\mathbf{r} - \frac{e}{2c}\mathbf{A}_r\right)^2|\psi\rangle, \quad \bar{U} = \langle\psi|U(r)|\psi\rangle,$$

$$\bar{\Pi} = \frac{2e^2}{M_e c^2}\langle\phi|\mathbf{A}_R^2|\phi\rangle, \qquad \hbar\tilde{\omega}_k = \hbar\omega_k^0 + \frac{2\hbar e}{M_e c}\langle\phi|\mathbf{k}\mathbf{A}_R|\phi\rangle. \tag{4.2.11}$$

In what follows in this section we will assume $\hbar = 1$, $\omega_k^0 = \omega_0 = 1$, $M_e = 1$. It follows from (4.2.10) that the difference between the bipolaron Hamiltonian and the polaron one is that in the latter V_k is replaced by $\bar{V}_k$ and $\bar{T}$, $\bar{U}$, $\bar{\Pi}$ are added to the polaron Hamiltonian.

With the use of the Lee–Low–Pines canonical transformation:

$$S_2 = \exp\left\{\sum_k f_k(a_k^+ - a_k)\right\}, \tag{4.2.12}$$

where f_k are variational parameters which stand for the value of the displacement of the field oscillators from their equilibrium positions:

$$S_2^{-1}a_kS_2 = a_k + f_k, \qquad S_2^{-1}a_k^+S_2 = a_k^+ + f_k, \tag{4.2.13}$$

Hamiltonian $\tilde{\tilde{H}}$:

$$\tilde{\tilde{H}} = S_2^{-1}\bar{\tilde{H}}S_2 , \qquad \tilde{\tilde{H}} = H_0 + H_1 , \tag{4.2.14}$$

will be written as

$$\begin{gathered} H_0 = 2\sum_k \bar{V}_k f_k + \sum_k f_k^2 \tilde{\omega}_k + \frac{1}{2}\left(\sum_k \mathbf{k} f_k^2\right)^2 + H_{KB} + \bar{T} + \bar{U} + \bar{\Pi}, \\ H_{KB} = \sum_k \omega_k a_k^+ a_k + \frac{1}{2}\sum_{k,k'} \mathbf{k}\mathbf{k}' f_k f_{k'}\left(a_k a_{k'} + a_k^+ a_{k'}^+ + a_k^+ a_{k'} + a_k a_{k'}^+\right), \end{gathered} \tag{4.2.15}$$

where

$$\omega_k = \tilde{\omega}_k + \frac{k^2}{2} + \mathbf{k}\sum_{k'} \mathbf{k}' f_{k'}^2 . \tag{4.2.16}$$

Hamiltonian H_1 contains the terms that are linear, threefold, and fourfold in the creation and annihilation operators. Its explicit form is given in Lakhno (2015b) and Tulub (1961) (see Appendix A).

Then, according to Lakhno (2015b) and Tulub (1961), the Bogolyubov–Tyablikov canonical transformation (Tyablikov, 1967) is used to pass on from the operators a_k^+, a_k to α_k^+, α_k:

$$a_k = \sum_{k'} M_{1kk'}\alpha_{k'} + \sum_{k'} M_{2kk'}^{*}\alpha_{k'}^+ , \qquad a_k^+ = \sum_{k'} M_{1kk'}^{*}\alpha_{k'}^+ + \sum_{k'} M_{2kk'}\alpha_{k'}, \tag{4.2.17}$$

where H_{KB} is a diagonal operator which makes vanish expectation H_1 in the absence of an external magnetic field (see Appendix A). The contribution of H_1 into the spectrum of the transformed Hamiltonian when the magnetic field is nonzero is discussed in Section 4.3.

In the new operators α_k^+, α_k Hamiltonian (4.2.15) takes on the form:

$$\begin{gathered} \tilde{\tilde{\tilde{H}}} = E_{bp} + \sum_k \nu_k \alpha_k^+ \alpha_k , \\ E_{bp} = \Delta E_r + 2\sum_k \bar{V}_k f_k + \sum_k \tilde{\omega}_k f_k^2 + \bar{T} + \bar{U} + \bar{\Pi}, \end{gathered} \tag{4.2.18}$$

where ΔE_r is the so-called recoil energy. A general expression for $\Delta E_r = \Delta E_r\{f_k\}$ was obtained in Section 2.3 (Tulub, 1961). The ground state energy E_{bp} was calculated in Chapter 2 (Lakhno, 2010b, 2012a; Kashirina et al., 2012) by minimization of (4.2.18) with respect to f_k and ψ in the absence of a magnetic field.

It should be noted that in the polaron theory with a broken symmetry a diagonal electron–phonon Hamiltonian takes the form (4.2.18) (Miyake, 1994). This Hamiltonian can be interpreted as a Hamiltonian of a polaron and a system of its associated renormalized actual phonons or a Hamiltonian which possesses a spectrum of quasiparticle excitations determined by (4.2.18) (Levinson and Rashba, 1974). In the latter case, the polaron excited states are Fermi quasiparticles.

In the case of a bipolaron, the situation is qualitatively different because a bipolaron is a Bose particle whose spectrum is determined by (4.2.18). Obviously, a gas of such particles can experience Bose–Einstein condensation (BEC). Treatment of (4.2.18) as a bipolaron and its associated renormalized phonons does not prevent their BEC, since maintenance of the particles required for BEC is fulfilled automatically since the total number of the renormalized phonons commutate with Hamiltonian (4.2.18).

Renormalized frequencies v_k involved in (4.2.18) according to (2.3.15) of Chapter 2 are determined by a secular equation for s:

$$1 = \frac{2}{3}\sum_k \frac{k^2 f_k^2 \omega_k}{s - \omega_k^2}, \tag{4.2.19}$$

whose solutions yield a spectrum of the values of $s = \{v_k^2\}$.

4.3 Energy spectrum of a TI bipolaron

Hamiltonian (4.2.18) can be conveniently presented as

$$\tilde{\tilde{\tilde{H}}} = \sum_{n=0,1,2\ldots} E_n \alpha_n^+ \alpha_n, \tag{4.3.1}$$

$$E_n = \begin{cases} E_{bp}, & n = 0; \\ E_{bp} + \omega_{k_n}, & n \neq 0; \end{cases} \tag{4.3.2}$$

where, in the case of a three-dimensional ionic crystal $\mathbf{k}_n$ is a vector with the components:

$$k_{n_i} = \pm \frac{2\pi(n_i - 1)}{N_{a_i}}, \quad n_i = 1, 2, \ldots, \frac{N_{a_i}}{2} + 1, \quad i = x, y, z, \tag{4.3.3}$$

N_{a_i} is the number of atoms along the ith crystallographic axis. Let us prove the validity of the expression for the spectrum (4.3.1) and (4.3.2). Since the operators α_n^+, α_n obey Bose commutation relations:

$$\left[\alpha_n, \alpha_{n'}^+\right] = \alpha_n \alpha_{n'}^+ - \alpha_{n'}^+ \alpha_n = \delta_{n,n'} \tag{4.3.4}$$

they can be considered to be operators of the birth and annihilation of TI bipolarons. The energy spectrum of TI bipolarons, according to (4.2.19), is given by the equation:

$$F(s) = 1 \tag{4.3.5}$$

where

$$F(s) = \frac{2}{3}\sum_n \frac{k_n^2 f_{k_n}^2 \omega_{k_n}^2}{s - \omega_{k_n}^2}.$$

It is convenient to solve equation (4.3.5) graphically (Fig. 4.1).

Figure 4.1 suggests that the frequencies ν_{k_n} lie between the frequencies ω_{k_n} и $\omega_{k_{n+1}}$. Hence, the spectrum ν_{k_n} as well as the spectrum ω_{k_n} is quasicontinuous: $\nu_{k_n} - \omega_{k_n} = 0(N^{-1})$, which just proves the validity of (4.3.1) and (4.3.2).

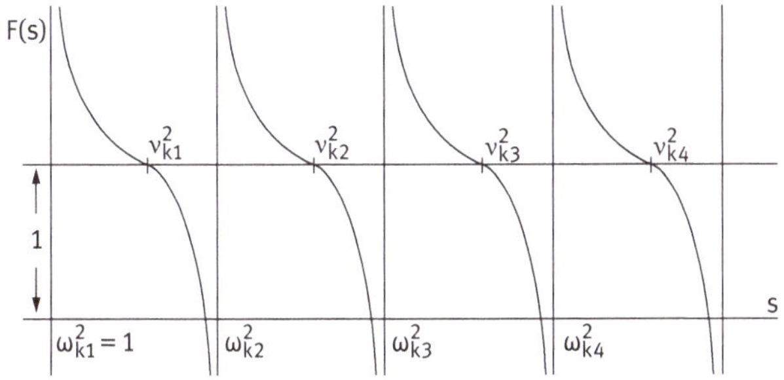

Fig. 4.1: Graphical solution of equation (4.3.5).

It follows that the spectrum of a TI bipolaron has a gap between the ground state E_{bp} and a quasicontinuous spectrum equal to ω_0.

In the absence of a magnetic field functions f_k involved in the expression for ω_k (4.2.16) are independent of the direction of the wave vector **k**. If a magnetic field is present f_k cannot be considered to be an isotropic value, accordingly, the last term in expression (4.2.16) for ω_k besides, an angular dependence involved the spectrum ω_k in a magnetic field is also contained in the term $\tilde{\omega}_k$involved in ω_k. Since in the isotropic system under consideration there is only one preferred direction determined by vector **B**, for ω_k from (4.2.16) we get:

$$\omega_{k_n} = \omega_0 + \frac{\hbar^2 k_n^2}{2M_e} + \frac{\eta}{M_e}(\mathbf{B}\mathbf{k}_n), \tag{4.3.6}$$

where η is a certain scalar value. It should be noted that the contribution of H_1 into the spectrum (4.3.6) leads to a dependence of η on $|\mathbf{k}|$ and $(\mathbf{kB})$ too. For a weak magnetic field in a long-wave limit (when Fröhlich Hamiltonian is applicable) we can neglect this dependence and believe that η is constant.

For a magnetic field directed along the z-axis, expression (4.3.6) can be presented in the form:

$$\omega_{k_n} = \omega_0 + \frac{\hbar^2}{2M_e}\left(k_{zn} + k_z^0\right)^2 + \frac{\hbar^2}{2M_e}\left(k_{xn}^2 + k_{yn}^2\right) - \frac{\eta^2 B^2}{2\hbar^2 M_e}, \tag{4.3.7}$$

It should be noted that Formula (4.3.7) can be generalized to the anisotropic case (which will be actual in what follows) when in the directions k_x and k_y: $M_{ex} = M_{ey} = M_{||}$, and in the direction k_z: $M_{ez} = M_{\perp}$ (Sections 4.7 and 4.8).

In this case, formula (4.3.7) takes the form:

$$\omega_{k_n} = \omega_0 + \frac{\hbar^2}{2M_{\perp}}\left(k_{zn} + k_z^0\right)^2 + \frac{\hbar^2}{2M_{||}}\left(k_{xn}^2 + k_{yn}^2\right) - \frac{\eta^2 B^2}{2\hbar^2 M_{\perp}}, \tag{4.3.7'}$$

if a magnetic field is directed along the z-axis and:

$$\omega_{k_n} = \omega_0 + \frac{\hbar^2}{2M_{\perp}}k_{zn}^2 + \frac{\hbar^2}{2M_{||}}\left(k_{xn} + k_x^0\right)^2 + \frac{\hbar^2}{2M_{||}}k_{yn}^2 - \frac{\eta^2 B^2}{2\hbar^2 M_{||}}, \tag{4.3.7''}$$

if a magnetic field is directed along the x-axis.

Below we consider the case of a low concentration of TI bipolarons in a crystal. In this case, as we will show in the next section, they can be considered as an ideal Bose gas whose properties are determined by Hamiltonian (4.3.1).

4.4 Nonideal gas of TI bipolarons

Being charged, a gas of TI bipolarons cannot be ideal since there should be a Coulomb interaction between the polarons. The theory of a nonideal gas implies that consideration of an interaction between the particles leads to qualitative changes in its spectral properties. According to Bogolyubov (1947), consideration of even a short-range interaction leads to a gap in the spectrum which is lacking in an ideal gas. Even more considerable changes can be expected in the case of a long-range Coulomb interaction. In this section we restrict ourselves by a lack of a magnetic field.

The logical scheme of the approach is as follows:

a) First we consider a particular case when there are only two electrons interacting with a phonon field. This is a classical bipolaron problem (Lakhno, 2015b).
b) Then we deal with a multiparticle problem that leads to the Fermi liquid concept. For this multielectron problem, we consider the case of two additional

electrons occurring above the Fermi surface (in its vicinity) bound by EPI (Cooper problem) (Lakhno, 2016b).

c) Then we believe that nearly all the electrons lying in the energy level within the layer $[E_F + E_{pol}, E_F]$, where E_F is the Fermi energy, E_{pol} is the polaron energy, occur in the TI polaron state; accordingly all the electrons in the narrow layer $[E_F + E_{bp}/2 - \delta E, E_F + E_{bp}/2 + \delta E]$, $\delta E \to 0$ occur in the TI bipolaron state, where E_{bp} is the energy of a TI bipolaron. A condensed bipolaron gas leads to an infinite density of electron states in this level.
d) Bipolarons are considered as charged bosons placed in an electron Fermi liquid (polaron gas) which screens an interaction between the bipolarons and the problem is reduced to that of a nonideal charged Bose gas.
e) The spectrum obtained is used to calculate the statistical properties of a TI bipolaron gas.

To develop a theory of a nonideal TI bipolaron gas we should know the spectrum of the states of an individual TI bipolaron in a polar medium. This problem was considered in detail in Lakhno (2018, 2019a) (Section 4.5). As it is shown in Lakhno (2016b), this spectrum will be the same as that of TI bipolarons which emerge near the Fermi surface. Hence, TI bipolarons in the layer $[E_F + E_{bp}/2,\ E_F]$ can be considered as a TI bipolaron Bose gas occurring in a polaron gas (Lakhno, 2017). If we believe that TI bipolarons do not interact with one another, such a gas can be considered to be ideal. Its properties will be fully determined if we know the spectrum of an individual TI bipolaron.

In considering the theory of an ideal gas and superconductivity on the basis of Bose particles of TI bipolarons, the Coulomb interaction between the electrons is taken into account only for electron pairs, that is, when we deal with the problem of an individual bipolarons. Hamiltonian of such a system, according to Lakhno (2015b) and Lakhno (2019a) has the form:

$$H_0 = \sum_k \varepsilon_k \alpha_k^+ \alpha_k, \tag{4.4.1}$$

$$\varepsilon_k = E_{bp}\Delta_{k,0} + \left(\omega_0 + E_{bp} + k^2/2M_e\right)\left(1 - \Delta_{k,0}\right), \tag{4.4.2}$$

where α_k^+, α_k are the operators of the birth and annihilation of TI bipolarons: ε_k is the spectrum of TI bipolarons obtained in Section 4.3; $\omega_0(\mathbf{k}) = \omega_0$ is the energy of an optical phonon; $\Delta_{k,0} = 1$ for $k = 0$ and $\Delta_{k,0} = 0$ for $k \neq 0$. Expressions (4.4.1) and (4.4.2) can be rewritten as

$$H_0 = E_{bp}\alpha_0^+ \alpha_0 + \sum_k{}' \left(\omega_0 + E_{bp} + k^2/2M_e\right)\alpha_k^+ \alpha_k, \tag{4.4.3}$$

where the prime in the sum in the right-hand side of (4.4.3) means that the term with $k = 0$ is lacking in the sum. Extraction out of a term with $k = 0$ in (4.4.1) corresponds to the formation of a Bose condensate where

$$\alpha_0 = \sqrt{N_0}, \tag{4.4.4}$$

N_0 is the number of TI bipolarons in a condensed state. Hence, in the theory of an ideal TI bipolaron gas the first term is merely $E_{bp}N_0$. In constructing a theory of a nonideal TI bipolaron Bose gas we will proceed from the Hamiltonian:

$$H = E_{bp}N_0 + \sum_k{}' \left(\omega_0 + E_{bp}\right)\alpha_k^+\alpha_k + \sum_k{}' t_k\alpha_k^+\alpha_k + 1/2V\sum_k{}' V_k\alpha^+_{k''-k}\alpha^+_{k'+k}\alpha_{k''}\alpha_{k'},$$

$$t_k = k^2/2M_e, \tag{4.4.5}$$

where the last term responsible for bipolaron interaction is added to Hamiltonian H_0 (4.4.1), V_k is a matrix element of the bipolaron interaction. The last two terms in (4.4.5) exactly correspond to Hamiltonian of a charged Bose gas (Foldy, 1961). Following a standard procedure of resolving a Bose condensate we rewrite (4.4.5) into the Hamiltonian:

$$H = E_{bp}N_0 + \sum_k{}' (\omega_0 + E_{bp})\alpha_k^+\alpha_k + \sum_k{}' \left[(t_k + n_0 V_k)\alpha_k^+\alpha_k + 1/2 n_0 V_k\left(\alpha_k\alpha_{-k} + \alpha_k^+\alpha_{-k}^+\right)\right], \tag{4.4.6}$$

where $n_0 = N_0/V$ is the density of the particles in the Bose condensate.

Then with the use of Bogolyubov transformation:

$$\alpha_k = u_k b_k - v_k b^+_{-k},$$

$$u_k = \left[(t_k + n_0 V_k + \varepsilon_k)/2\varepsilon_k\right]^{1/2},$$

$$v_k = \left[(t_k + n_0 V_k - \varepsilon_k)/2\varepsilon_k\right]^{1/2}, \qquad \varepsilon_k = \left[2n_0 V_k t_k + t_k^2\right]^{1/2}, \tag{4.4.7}$$

In new operators we get the Hamiltonian:

$$H = E_{bp}N_0 + U_0 + \sum_k{}' \left(\omega_0 + E_{bp} + \varepsilon_k\right) b_k^+ b_k, \tag{4.4.8}$$

$$U_0 = \sum_k{}' (\varepsilon_k - t_k - n_0 V_k),$$

where U_0 is the ground state energy of a charged Bose gas without regard to its interaction with the crystal polarization. Hence, the excitation spectrum of a nonideal TI bipolaron gas has the form:

$$E_k = E_{bp} + u_0 + \left(\omega_0(\mathbf{k}) + \sqrt{k^4/4M_e^2 + k^2V_kn_0/M_e}\right)(1-\Delta_{k,0}), \tag{4.4.9}$$

where $u_0 = U_n/N$, N is the total number of particles. If we reckon the excitation energy from the bipolaron ground state energy in a nonideal gas, assuming that $\Delta_k = E_k - (E_{bp} + u_0)$, then for Δ_k (when $k \neq 0$) we get:

$$\Delta_k = \omega_0(\mathbf{k}) + \sqrt{k^4/4M_e^2 + k^2V_kn_0/M_e}. \tag{4.4.10}$$

This spectrum suggests that a TI bipolaron gas has a gap Δ_k in the spectrum between the ground and excited states, that is, superfluid. Being charged, this gas is automatically superconducting. To determine a particular form of the spectrum we should know the value of V_k. If we considered only a charged Bose gas with a positive homogeneous background, produced by a rigid ion lattice, then the quantity $V_k$involved in (4.4.9) in the absence of screening would be equal to $V_k = 4\pi e_B^2/k^2$. Accordingly, the second term in the radical expression in (4.4.9) would be equal to $\omega_p^2 = 4\pi n_0 e_B^2/M_e$, where ω_p is the plasma frequency of a Boson gas, e_B is the Boson charge ($2e$ for a TI bipolaron). Actually, if screening is taken into account, V_k takes the form $V_k = 4\pi e^2/k^2\varepsilon_B(k)$, where $\varepsilon_B(k)$ is the dielectric permittivity of a charged Bose gas which was calculated in Hore and Frankel (1975, 1976). The expression for $\varepsilon_B(k)$ obtained in Hore and Frankel (1975, 1976) is too complicated and is not presented here. However, in the case of a TI bipolaron Bose gas this modification of V_k is insufficient. As it was shown in Lakhno (2016a, 2016b) (Chapter 5), bipolarons constitute just a small portion of charged particles in the system. Most of them occur in the electron gas into which the bipolarons are immersed. It is just the electron gas that makes the main contribution into the screening of the interaction between the polarons. To take account of this screening V_k should be expressed as $V_k = 4\pi e^2/k^2\varepsilon_B(k)\varepsilon_e(k)$, where $\varepsilon_e(k)$ is the dielectric permittivity of an electron gas (Appendix H). Finally, if we take account of the mobility of the ion lattice, V_k takes the form: $V_k = 4\pi e^2/k^2\varepsilon_B(k)\varepsilon_e(k)\varepsilon_\infty\varepsilon_0$, where ε_∞, ε_0 are the high-frequency and static dielectric constants.

As a result, we get for Δ_k:

$$\Delta_k = \omega_0(\mathbf{k}) + k^2/2M_e\sqrt{1+\chi(k)}, \tag{4.4.11}$$

$$\chi(k) = (2M_e\omega_p)^2/k^4\varepsilon_B(k)\varepsilon_e(k)\varepsilon_\infty\varepsilon_0. \tag{4.4.12}$$

To estimate $\chi(k)$ in (4.4.11) let us consider the long-wave limit. In this limit $\varepsilon_e(k)$ has the Thomas–Fermi form: $\varepsilon_e(k) = 1 + \kappa^2/k^2$, where $\kappa = 0.815\, k_F(r_s/a_B)^{1/2}$, $a_B = \hbar^2/M_e e_B^2$, $r_s = (3/4\pi n_0)^{1/3}$, therefore, according to Hore and Frankel (1975, 1976), the quantity $\varepsilon_B(k)$ is equal to $\varepsilon_B(k) = 1 + q_s^4/k^4$, $q_s = \sqrt{2M_e\omega_p}$.

Bearing in mind that in calculations of the thermodynamic functions the main contribution is made by the values of k: $k^2/2M_e \approx T$, where T is the temperature for

$\chi(k)$ we get an estimate $\chi \sim T/E_F\varepsilon_\infty\varepsilon_0$ where E_F is the Fermi energy. Hence the spectrum of a screened TI bipolaron gas differs from the spectrum of an ideal TI bipolaron gas (4.4.2) only slightly. It should be noted that in view of screening the value of correlation energy u_0 in (4.4.10) turns out to be much less than that calculated in Foldy (1961) without screening and for actual parameter values – much less than the bipolaron energy $|E_{bp}|$. It should also be noted that in view of screening a TI bipolaron gas does not form Wigner crystal even in the case of an arbitrarily small bipolaron density.

4.5 Statistical thermodynamics of a low-density TI bipolaron gas

In accordance with the result of the previous section let us consider an ideal Bose gas of TI bipolarons which is a system of N particles occurring in a volume V (Lakhno, 2018, 2019b). Let us write N_0 for the number of particles in the lower one-particle state, and N' – for the number of particles in higher states. Then:

$$N = \sum_{n=0,1,2,\ldots} \bar{m}_n = \sum_n \frac{1}{e^{(E_n-\mu)/T}-1}, \tag{4.5.1}$$

or:

$$N = N_0 + N', \quad N_0 = \frac{1}{e^{(E_{bp}-\mu)/T}-1}, \quad N' = \sum_{n\neq 0} \frac{1}{e^{(E_n-\mu)/T}-1}. \tag{4.5.2}$$

If in the expression for N' (4.5.2) we replace summation by integration over a continuous spectrum (4.3.1), (4.3.2), and (4.3.7) and assume in (4.5.2) that $\mu = E_{bp}$ we will get from (4.5.1) and (4.5.2) an equation for the critical temperature of Bose condensation T_c:

$$C_{bp} = f_{\tilde{\omega}_H}(\tilde{T}_c), \tag{4.5.3}$$

$$f_{\tilde{\omega}_H}(\tilde{T}_c) = \tilde{T}_c^{3/2} F_{3/2}(\tilde{\omega}_H/\tilde{T}_c), \quad F_{3/2}(\alpha) = \frac{2}{\sqrt{\pi}} \int_0^\infty \frac{x^{1/2}dx}{e^{x+\alpha}-1}$$

$$C_{bp} = \left(\frac{n^{2/3} 2\pi\hbar^2}{M_e \omega *}\right)^{3/2}, \quad \tilde{\omega}_H = \frac{\omega_0 - \eta^2 H^2/2M_e}{\omega *}, \quad \tilde{T}_c = \frac{T_c}{\omega *}$$

where $n = N/V$. In this section we deal with the case when a magnetic field is absent: $H = 0$. Figure 4.2 shows a graphical solution of equation (4.5.3) for the parameter values $M_e = 2m = 2m_0$, where m_0 – is a mass of a free electron in vacuum, $\omega^* = 5$ meV(≈ 58 K), $n = 10^{21}\text{cm}^{-3}$ and the values: $\tilde{\omega}_1 = 0.2$; $\tilde{\omega}_2 = 1$; $\tilde{\omega}_3 = 2$; $\tilde{\omega}_4 = 10$; $\tilde{\omega}_5 = 15$; $\tilde{\omega}_6 = 20$; $\tilde{\omega}_H = \tilde{\omega} = \tilde{\omega}_0/\omega^*$; $\omega_{H_{cr}} = \omega_H$, for $\tilde{T} = \tilde{T}_c$.

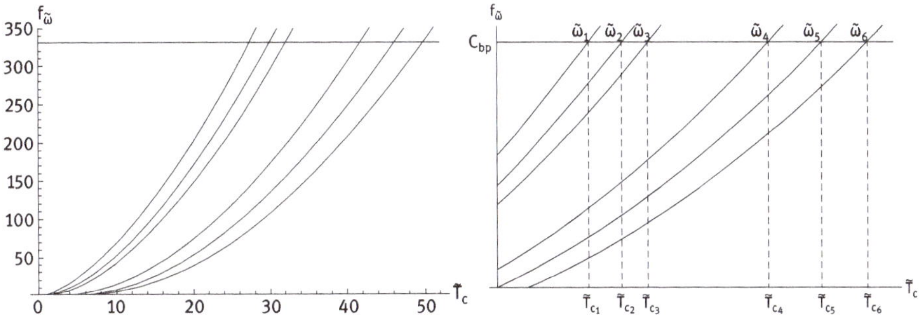

Fig. 4.2: Solution of equation (4.5.3) with $C_{bp} = 331.3$ and $\tilde{\omega}_i = \{0.2; 1; 2; 10; 15; 20\}$, which correspond to $\tilde{T}_{ci}$: $\tilde{T}_{c_1} = 27.3$; $\tilde{T}_{c_2} = 30$; $\tilde{T}_{c_3} = 32$; $\tilde{T}_{c_4} = 42$; $\tilde{T}_{c_5} = 46.2$; $\tilde{T}_{c_6} = 50$.

Figure 4.2 suggests that the critical temperature grows as the phonon frequency $\tilde{\omega}_0$ increases. Relations of critical temperatures T_{ci}/ω_{0i} corresponding to the chosen parameter values are given in Tab. 4.1. It can be seen from Tab. 4.1 that the critical temperature of a TI bipolaron gas is always higher than in the case of an ideal Bose gas (IBG). It also follows from Fig. 4.2 that as the concentration of TI bipolarons n increases the critical temperature will increase and as the electron mass m grows the critical temperature will decrease.

For $\tilde{\omega} = 0$, the results pass on to the well-known IBG limit. In particular, from (4.5.3) for $\tilde{\omega} = 0$ follows the expression for the critical temperature of IBG:

$$T_c = 3.31\, \hbar^2 n^{2/3} / M_e. \tag{4.5.4}$$

It should be stressed that (4.5.4) involves $M_e = 2m$, rather than the bipolaron mass. This eliminates the problem of low condensation temperature which arises both in the SRP and LRP theories where expression (4.5.4) involves the bipolaron mass (Alexandrov and Krebs, 1992; Alexandrov and Mott, 1996; Ogg, 1946; Vinetsky and Pashitsky, 1975; Vinetsky et al.,1989; Vinetsky et al., 1992; Pashitskii and Vineckii, 1987; Emin, 1989, 2017). Another important result is that the critical temperature T_c for the parameter values chosen is considerably superior to the gap energy ω_0.

From (4.5.1) and (4.5.2), it follows that:

$$\frac{N'(\tilde{\omega})}{N} = \frac{\tilde{T}^{3/2}}{C_{bp}} F_{3/2}\left(\frac{\tilde{\omega}}{\tilde{T}}\right), \tag{4.5.5}$$

$$\frac{N_0(\tilde{\omega})}{N} = 1 - \frac{N'(\tilde{\omega})}{N}. \tag{4.5.6}$$

Tab. 4.1: Calculated values of characteristics of a TI bipolaron Bose gas with concentration $n = 10^{21}\text{cm}^{-3}$ $\tilde{\omega}_i = \omega_i/\omega^*$, $\omega^* = 5\text{meV}$ (≈ 58 K), ω_i is the energy of an optical phonon; T_{c_i} a critical temperature of the transition, q_i is a latent heat of the transition from the condensate to supracondensate state; $-\Delta(\partial C_{vi}/\partial\tilde{T}) = \partial C_{vi}/\partial\tilde{T}|_{\tilde{T}=T_{ci}+0} - \partial C_{vi}/\partial\tilde{T}|_{\tilde{T}=T_{ci}-0}$ is a jump of the heat capacity under transition, $\tilde{T} = T/\omega^*$; $C_{vi}(T_c - 0)$ is the heat capacity in the SC phase at the critical point; $C_s = C_v(T_c - 0)$, $C_n = C_v(T_c + 0)$. Calculations are performed for the concentration of TI bipolarons $n = 10^{21}$ cm $^{-3}$ and the effective mass of a band electron $m = m_0$. The table also lists the values of the concentrations of TI bipolarons n_{bpi} for HTSC $YBa_2Cu_3O_7$, proceeding from the experimental value of the transition temperature $T_c = 93$ K.

i	0	1	2	3	4	5	6
$\tilde{\omega}_i$	0	0.2	1	2	10	15	20
T_{ci}/ω_{0i}	∞	136.6	30	16	4.2	3	2.5
q_i/T_{ci}	1.3	1.44	1.64	1.8	2.5	2.8	3
$-\Delta(\partial C_{vi}/\partial\tilde{T})$	0.11	0.12	0.12	0.13	0.14	0.15	0.15
$C_{vi}(T_c - 0)$	1.9	2.16	2.46	2.7	3.74	4.2	1,6
$(C_s - C_n)/C_n$	0	0.16	0.36	0.52	1.23	1.53	1.8
$n_{bpi} \times \text{cm}^3$	16×10^{19}	9.4×10^{18}	4.2×10^{18}	2.0×10^{18}	1.2×10^{17}	5.2×10^{14}	2.3×10^{13}

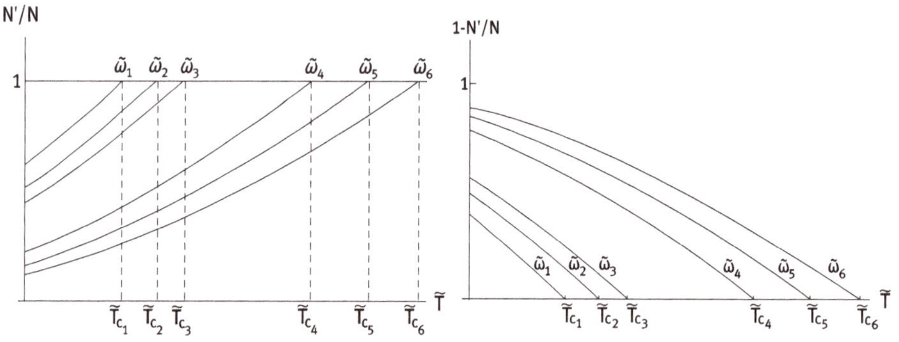

Fig. 4.3: Temperature dependencies of the relative number of supracondensate particles N'/N and particles in the condensate $N_0/N = 1 - N'/N$ for the values of parameters $\tilde{\omega}_i$ given in Fig. 4.2.

Figure 4.3 shows the temperature dependencies of the number of supracondensate N' and condensate N_0 particles for the above indicated values of $\tilde{\omega}_i$.

Figure 4.3 suggests that, as one would expect, the number of condensate particles grows as the gap $\tilde{\omega}_i$ increases.

The energy of a TI bipolaron gas E is determined by the expression:

$$E = \sum_{n=0,1,2\ldots} \bar{m}_n E_n = E_{bp} N_0 + \sum_{n \neq 0} \bar{m}_n E_n. \tag{4.5.7}$$

With the use of (4.3.1), (4.3.2), and (4.5.7), we express the specific energy (i.e., energy per one TI bipolaron) $\tilde{E}(\tilde{T}) = E/N\omega^*$, $\tilde{E}_{bp} = E_{bp}/\omega^*$ as

$$\tilde{E}(\tilde{T}) = \tilde{E}_{bp} + \frac{\tilde{T}^{5/2}}{C_{bp}} F_{3/2}\left(\frac{\tilde{\omega}-\tilde{\mu}}{\tilde{T}}\right)\left[\frac{\tilde{\omega}}{\tilde{T}} + \frac{F_{5/2}\left(\frac{\tilde{\omega}-\tilde{\mu}}{\tilde{T}}\right)}{F_{3/2}\left(\frac{\tilde{\omega}-\tilde{\mu}}{\tilde{T}}\right)}\right],$$

$$F_{5/2}(\alpha) = \frac{2}{\sqrt{\pi}} \int_0^\infty \frac{x^{3/2} dx}{e^{x+\alpha} - 1} \tag{4.5.8}$$

where $\tilde{\mu}$ is determined by the equation:

$$\tilde{T}^{3/2} F_{3/2}\left(\frac{\tilde{\omega}-\tilde{\mu}}{\tilde{T}}\right) = C_{bp}, \qquad \tilde{\mu} = \begin{cases} 0, & \tilde{T} \leq \tilde{T}_c \\ \tilde{\mu}(\tilde{T}), & \tilde{T} \geq \tilde{T}_c \end{cases}. \tag{4.5.9}$$

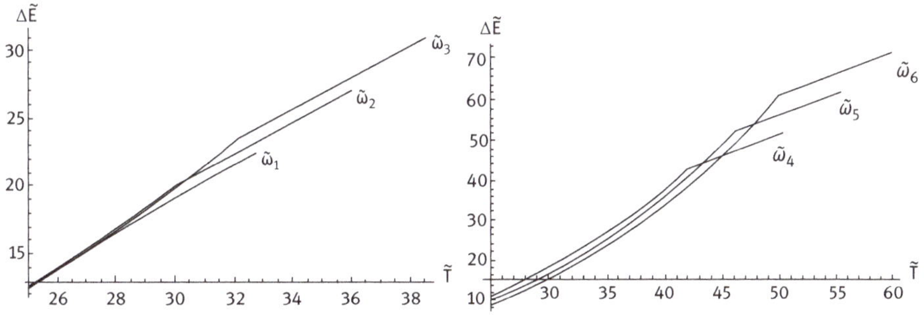

Fig. 4.4: Temperature dependencies of $\Delta\tilde{E}(\tilde{T}) = \tilde{E}(\tilde{T}) - \tilde{E}_{bp}$ for $\tilde{\omega}_i$ values presented in Figs. 4.2 and 4.3.

A relation of $\tilde{\mu}$ to the chemical potential of the system μ is determined by the expression $\tilde{\mu} = (\mu - E_{bp})/\omega^*$. Formulae (4.5.8)–(4.5.9) also yield the expressions for the free energy: $\Delta F = -2\Delta E/3$, $\Delta F = F(T) - F(0)$, $\Delta E = E(T) - E(0)$, and entropy $S = -\partial F/\partial T$.

Figure 4.4 shows the temperature dependencies of $\Delta\tilde{E} = \tilde{E} - \tilde{E}_{bp}$ for the aforementioned values of $\tilde{\omega}_i$. Breakpoints of the $\Delta\tilde{E}_i(\tilde{T})$ curves correspond to the critical temperature values T_{ci}.

The dependencies obtained enable us to find the heat capacity of a TI bipolaron gas: $C_v(\tilde{T}) = d\tilde{E}/d\tilde{T}$. With the use of (4.5.8) for $\tilde{T} \le \tilde{T}_c$ we express $C_v(\tilde{T})$ as

$$C_v(\tilde{T}) = \frac{\tilde{T}^{3/2}}{2C_{bp}}\left[\frac{\tilde{\omega}^2}{\tilde{T}^2}F_{1/2}\left(\frac{\tilde{\omega}}{\tilde{T}}\right) + 6\left(\frac{\tilde{\omega}}{\tilde{T}}\right)F_{3/2}\left(\frac{\tilde{\omega}}{\tilde{T}}\right) + 5F_{5/2}\left(\frac{\tilde{\omega}}{\tilde{T}}\right)\right], \tag{4.5.10}$$

$$F_{1/2}(\alpha) = \frac{2}{\sqrt{\pi}}\int_0^\infty \frac{1}{\sqrt{x}}\frac{dx}{e^{x+\alpha}-1}.$$

Expression (4.5.10) yields a well-known exponential dependence of the heat capacity at low temperatures $C_v \sim \exp(-\omega_0/T)$, caused by the availability of the energy gap ω_0.

Figure 4.5 illustrates the temperature dependencies of the heat capacity $C_v(\tilde{T})$ for the above-mentioned values of $\tilde{\omega}_i$. Table 4.1 lists the values of jumps in the heat capacity for different $\tilde{\omega}_i$:

$$\Delta\frac{\partial C_v(\tilde{T})}{\partial \tilde{T}} = \left.\frac{\partial C_v(\tilde{T})}{\partial \tilde{T}}\right|_{\tilde{T}=T_C+0} - \left.\frac{\partial C_v(\tilde{T})}{\partial \tilde{T}}\right|_{\tilde{T}=T_C-0} \tag{4.5.11}$$

at the transition points.

The dependencies obtained enable one to find a latent transition heat $q = TS$, S is the entropy of supracondensate particles. At the transition point this value is equal to $q = 2T_cC_v(T_c-0)/3$, where $C_v(T)$ is determined by formula (4.5.10), and for the abovementioned values of $\tilde{\omega}_i$ is given in Tab. 4.1.

Above, we considered the thermodynamic characteristics of a Bose condensate with an isotropic phonon spectrum. In fact, in most HTSC materials, the SC gap depends on the wave vector. For example, in YBCO with optimal doping, the dependence of the gap on **k** has the form:

$$\omega_0(\mathbf{k}) = \Delta_0\left|\cos k_xa - \cos k_ya\right| + \omega_0, \tag{4.5.12}$$

where $\omega_0(\mathbf{k})$, according to the above, is the phonon frequency. In TI bipolaron theory, the first term on the right-hand side of (4.5.12) corresponds to the contribution of the d-type wave, and the second term to the contribution of the s-type wave.

When calculating the thermodynamic properties of the gas of TI bipolarons, the quantity $\omega_0(\mathbf{k})$ enters into the expression for the spectrum ν_k in the form:

$$\nu_k = E_{bp}\Delta_{k,0} + \left(E_{bp} + \omega_0(\mathbf{k}) + k^2/2M\right)(1-\Delta_{k,0}). \tag{4.5.13}$$

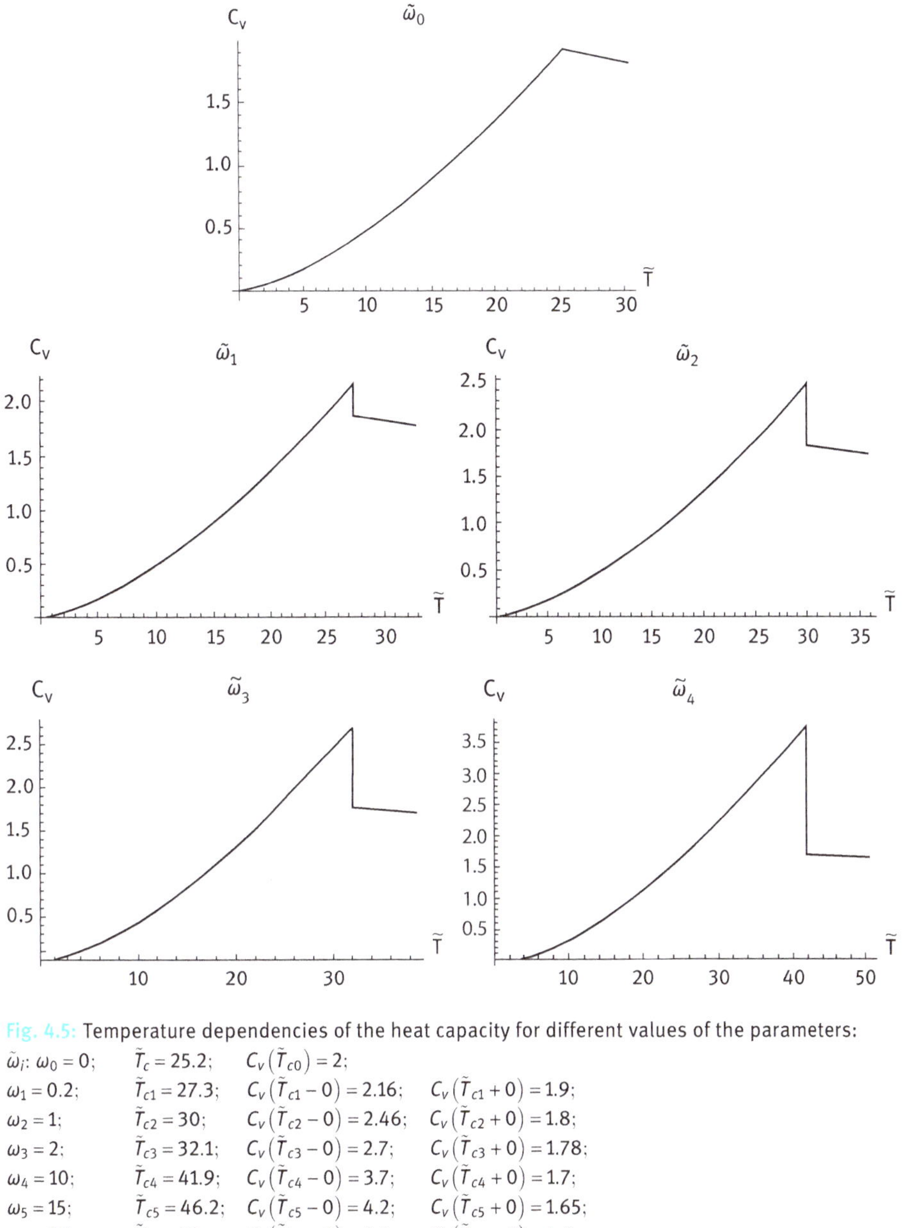

Fig. 4.5: Temperature dependencies of the heat capacity for different values of the parameters:

$\tilde{\omega}_i$: $\omega_0 = 0$;	$\tilde{T}_c = 25.2$;	$C_v\left(\tilde{T}_{c0}\right) = 2$;	
$\omega_1 = 0.2$;	$\tilde{T}_{c1} = 27.3$;	$C_v\left(\tilde{T}_{c1} - 0\right) = 2.16$;	$C_v\left(\tilde{T}_{c1} + 0\right) = 1.9$;
$\omega_2 = 1$;	$\tilde{T}_{c2} = 30$;	$C_v\left(\tilde{T}_{c2} - 0\right) = 2.46$;	$C_v\left(\tilde{T}_{c2} + 0\right) = 1.8$;
$\omega_3 = 2$;	$\tilde{T}_{c3} = 32.1$;	$C_v\left(\tilde{T}_{c3} - 0\right) = 2.7$;	$C_v\left(\tilde{T}_{c3} + 0\right) = 1.78$;
$\omega_4 = 10$;	$\tilde{T}_{c4} = 41.9$;	$C_v\left(\tilde{T}_{c4} - 0\right) = 3.7$;	$C_v\left(\tilde{T}_{c4} + 0\right) = 1.7$;
$\omega_5 = 15$;	$\tilde{T}_{c5} = 46.2$;	$C_v\left(\tilde{T}_{c5} - 0\right) = 4.2$;	$C_v\left(\tilde{T}_{c5} + 0\right) = 1.65$;
$\omega_6 = 20$;	$\tilde{T}_{c6} = 50$	$C_v\left(\tilde{T}_{c6} - 0\right) = 4.6$;	$C_v\left(\tilde{T}_{c6} + 0\right) = 1.6$..

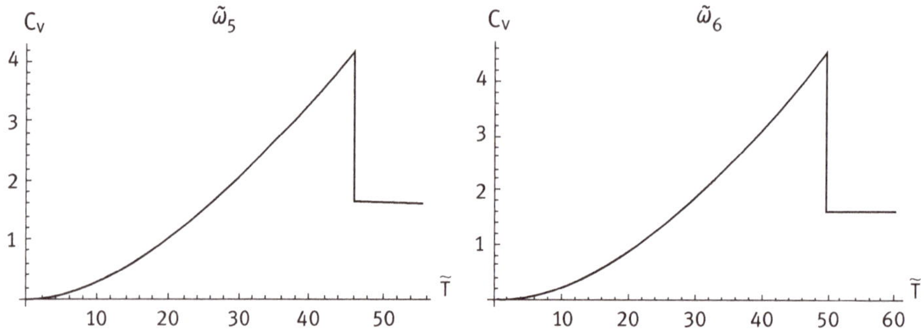

Fig. 4.5 (continued)

As a result, the quantity N', determined by (4.5.2):

$$N' = \frac{V}{(2\pi\hbar)^3}\int d^3k \frac{1}{e^{(\nu_k-\mu)/T}-1}, \tag{4.5.14}$$

can be estimated as follows. Equations (4.5.12) and (4.5.13) suggest the following condition when the main contribution into (4.5.14) is made by s-wave. Taking into account that the main contribution into integral (4.5.14) is made by the values of $k \approx \sqrt{2MT}$ and $ka \ll 1$ we get from (4.5.12) and (4.5.13) that $\Delta_0|\cos k_x a - \cos k_y a| \cong \Delta_0 Ma^2/\hbar^2 \ll 1$.

Hence, for $\omega_0 \geq \Delta_0 Ma^2T/\hbar^2$ the main contribution into the integral will be made by s-wave. In this case the results obtained above will remain unchanged. Thus, for example in the case of YBCO the values of ω_0/Δ_0 is ≈ 0.15 (Smilde et al., 2005; Kirtley et al., 2006) and the s-approximation condition in this case is satisfied with high accuracy.

4.6 Statistical thermodynamics of 1D low-density TI bipolaron gas

The spectrum of Holstein 1D bipolaron, as in 3D-case is determined by equation $D(s)=0$, where $D(s)$ is determined by equation (3.3.4) of Chapter 3 and has the form $E_n = E_{BP}$, $n=0$; $E_n = E_{BP} + \omega_0 + k_n^2/2$, $n \neq 0$, $k_n = \pm 2\pi(n-1)/N_a$, $n = 1, 2, \ldots, N_a/2+1$. With the use of (4.5.1), (4.5.2), and (4.5.3), we get an equation for the critical temperature of Bose condensation:

$$C_{1D} = \phi_{\tilde{\omega}}(T_C), \tag{4.6.1}$$

$$\phi_{\tilde{\omega}} = \tilde{T}_C^{1/2} F_{1/2}(\tilde{\omega}/\tilde{T}_C), \quad F_{1/2}(\alpha) = \int_0^{\infty} \frac{dx}{\sqrt{x}(e^{x+\alpha}-1)},$$

$$C_{1D} = 2\sqrt{2}\pi \frac{n\hbar}{M^{1/2}\omega^{*1/2}}, \quad \omega^* = \omega_0/\omega, \quad \tilde{T}_C = T/\omega^*,$$

where $n = N/L$ is the number of bipolarons per unit length. Figure 4.6 shows a graphical solution of equation (4.6.1) for the values of the parameters: $\omega^* = 5\text{meV}(\approx 58\ \text{K})$, $n = 10^7\text{cm}^{-1}$ and the values $\tilde{\omega}_1 = 0.2$; $\tilde{\omega}_2 = 1$; $\tilde{\omega}_3 = 2$; $\tilde{\omega}_4 = 10$; $\tilde{\omega}_5 = 15$; $\tilde{\omega}_6 = 20$; $C_{1D} = 34.69$.

Figure 4.6 suggests that the critical temperature grows as the phonon frequency grows and vanishes for $\omega = 0$. The equality $T_C = 0$ for $\omega = 0$ corresponds to a well-known result that Bose condensation is impossible in one-dimensional case.
Figure 4.6 also suggests that an increase in the concentration of bipolarons will lead to an increase in the critical temperature, while an increase in the electron effective mass m leads to its decrease.

It follows from (4.5.1) and (4.5.2) that:

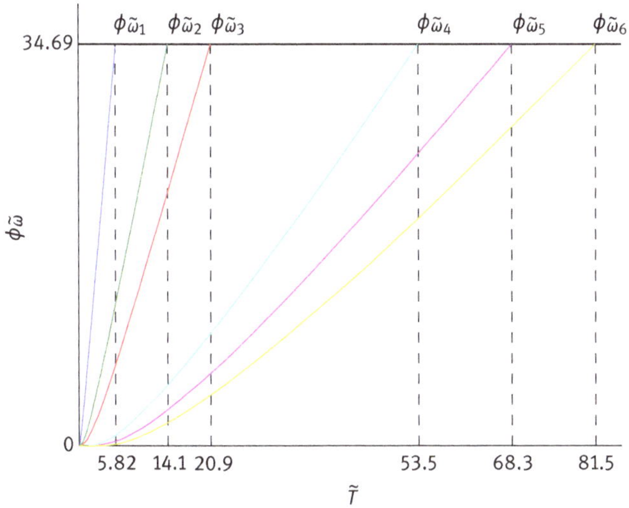

Fig. 4.6: Solutions of equation (4.6.1) with $C_{1D} = 34.69$ and $\tilde{\omega}_i = 0, 2; 1; 2; 10; 15; 20$, which correspond to $\tilde{T}_{c_i}$: $\tilde{T}_{c1} = 5, 8$; $\tilde{T}_{c2} = 14$; $\tilde{T}_{c3} = 20, 9$; $\tilde{T}_{c4} = 53, 5$; $\tilde{T}_{c5} = 68, 3$; $\tilde{T}_{c6} = 81, 5$.

$$\frac{N'(\tilde{\omega})}{N} = \frac{\tilde{T}^{1/2}}{C_{1D}} F_{1/2}\left(\frac{\tilde{\omega}}{\tilde{T}}\right) \tag{4.6.2}$$

$$\frac{N_0(\tilde{\omega})}{N} = 1 - \frac{N'(\tilde{\omega})}{N} \tag{4.6.3}$$

Figure 4.7 shows temperature dependencies of the number of supracondensate particles N' and the number of condensate particles N_0 for different values of the parameters $\tilde{\omega}_C = \tilde{\omega}(T_C)$.

Figure 4.7 suggests as one would expect that the number of particles in the condensate increases with increasing gap energy ω_i.

Using the above expression for the spectrum of 1D bipolaron we rewrite the expression for the specific energy of a bipolaron gas (4.5.7) as

$$\tilde{E} = \tilde{E}_{BP} + \Delta\tilde{E}, \quad \tilde{E}(\tilde{T}) = E/N\omega^*, \quad \tilde{E}_{BP} = E_{BP}/\omega^* \tag{4.6.4}$$

$$\Delta\tilde{E} = \frac{\tilde{T}^{3/2}}{C_{1D}} F_{1/2}\left(\frac{\tilde{\omega}-\tilde{\mu}}{\tilde{T}}\right)\left[\frac{\tilde{\omega}}{\tilde{T}} + \frac{F_{3/2}\left(\frac{\tilde{\omega}-\tilde{\mu}}{\tilde{T}}\right)}{F_{1/2}\left(\frac{\tilde{\omega}-\tilde{\mu}}{\tilde{T}}\right)}\right], \tag{4.6.5}$$

where the chemical potential μ is determined from the equation:

$$C_{1D} = \tilde{T}_C^{1/2} F_{1/2}\left(\frac{\tilde{\omega}-\tilde{\mu}(\tilde{T})}{\tilde{T}}\right), \ \tilde{\mu} = \begin{cases} 0, & \tilde{T} < \tilde{T}_C \\ \tilde{\mu}(\tilde{T}), & \tilde{T} > \tilde{T}_C. \end{cases} \tag{4.6.6}$$

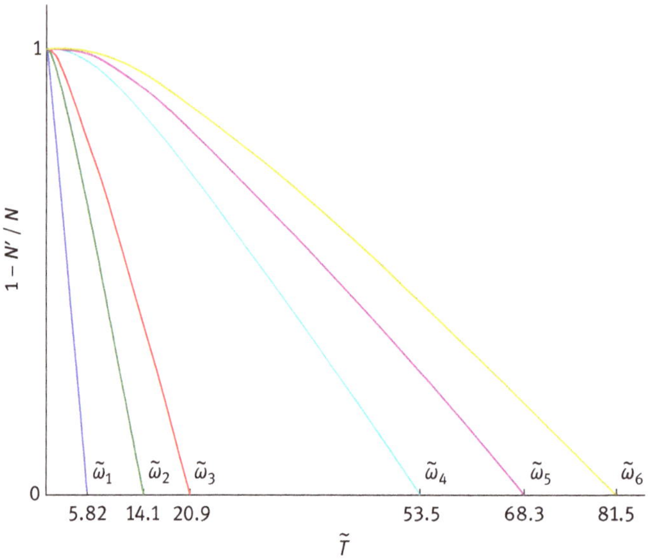

Fig. 4.7: Temperature dependencies of the relative number of supracondensate particles (accordingly condensed particles $N_0/N = 1 - N'/N$). For the values of the parameters $\tilde{\omega}_i$ given in Fig. 4.6.

A relation between $\tilde{\mu}$ and the chemical potential of the system μ is determined by the expression: $\tilde{\mu} = (\mu - E_{BP})/\omega^*$. Formulae (4.6.5) and (4.6.6) lead to the following expression for the thermodynamic potential Ω: $\Delta\Omega = -2\Delta E$, $\Delta\Omega = \Omega(T) - \Omega(0)$, $\Delta E = E(T) - E(0)$ and entropy $S = -\partial\Omega/\partial T$ $(F = -2E,\ S = -\partial F/\partial T)$.

Figure 4.8 shows temperature dependencies $\Delta\tilde{E} = \tilde{E} - \tilde{E}_{BP}$ for the above values of $\tilde{\omega}_i$. The salient point in the dependencies $\Delta\tilde{E}_i(\tilde{T})$ corresponds to the values of critical temperatures T_{c_i}.

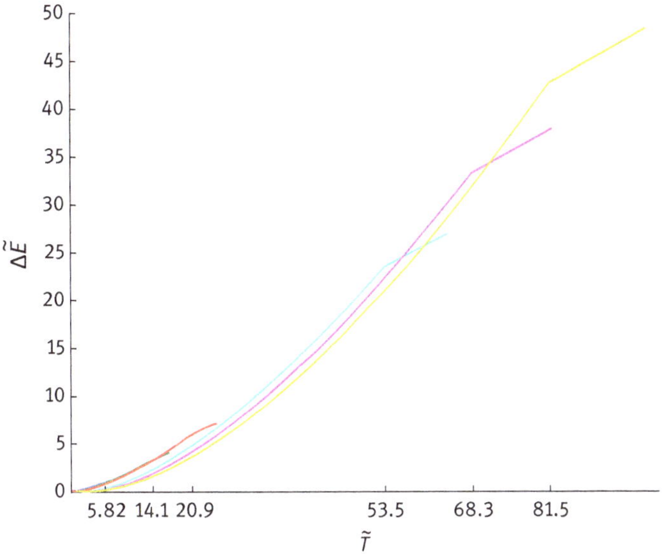

Fig. 4.8: Temperature dependencies $\Delta\tilde{E} = \tilde{E}(\tilde{T}) - \tilde{E}_{BP}$ for the parameter values of $\tilde{\omega}_i$, presented in Fig. 4.6.

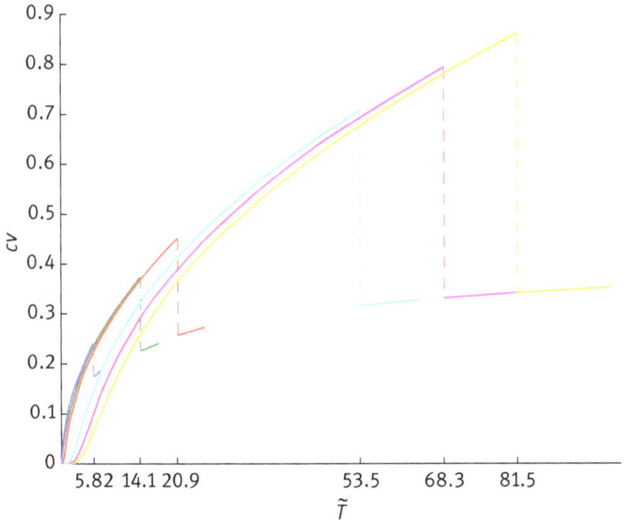

Fig. 4.9: Temperature dependencies of heat capacities for the parameter values of $\tilde{\omega}_i$ presented in Fig. 4.6 (numerical values of jumps in the heat capacity are presented in Tab. 4.2).

Tab. 4.2: Numerical values of the jumps in the heat capacity Δ (4.5.11) for the values of $\tilde{\omega}_i$ at the points of transition.

i	1	2	3	4	5	6
$\tilde{\omega}_i$	0.2	1	2	10	15	20
$\tilde{T}_{c_i}$	5.82	14.11	20.87	53.47	68.33	81.5
$C_V\left(\tilde{T}_{c_i}-0\right)$	0.24	0.37	0.45	0.71	0.79	0.86
$C_V\left(\tilde{T}_{c_i}+0\right)$	0.17	0.23	0.25	0.32	0.33	0.34
$\frac{\partial C_V}{\partial \tilde{T}}\left(\tilde{T}_{c_i}-0\right)$	20.88×10^{-3}	13.51×10^{-3}	11.16×10^{-3}	7.11×10^{-3}	6.33×10^{-3}	5.83×10^{-3}
$\frac{\partial C_V}{\partial \tilde{T}}\left(\tilde{T}_{c_i}+0\right)$	10.22×10^{-3}	4.72×10^{-3}	3.24×10^{-3}	1.19×10^{-3}	0.89×10^{-3}	0.73×10^{-3}
Δ	-10.66×10^{-3}	-8.79×10^{-3}	-7.93×10^{-3}	-5.92×10^{-3}	-5.44×10^{-3}	-5.11×10^{-3}

The dependencies obtained enable us to determine the heat capacity of 1D TI bipolaron gas: $C_V(\tilde{T}) = d\tilde{E}/d\tilde{T}$.

Figure 4.9 shows temperature dependencies $C_V(\tilde{T})$ for the values of $\tilde{\omega}_i$ presented above.

The dependencies obtained enable one to find the heat of transition $q = TS$, where S is the entropy of supracondensate particles. At the points of transition this value will be equal to $q = 2T_c \cdot C_V(T_c - 0) = d\tilde{E}/d\tilde{T}$, where $\tilde{E}$ is determined by formulae (4.6.4) and (4.6.5). The values of the heats of transition q_i for the values of $\tilde{\omega}_i$ presented above are given in Tab. 4.2.

4.7 Current states of a TI bipolaron gas

As is well-known, an absence of a magnetic field in a superconductor is caused by a presence of surface currents which compensate this field. Thus, it follows from condition (4.2.8) that:

$$\mathbf{P}_R = -\frac{2e}{c}\mathbf{A}_R, \tag{4.7.1}$$

that is, in the superconductor there is a persistent current $\mathbf{j}$:

$$\mathbf{j} = \frac{2en_0\mathbf{P}_R}{M_e^*} = -\frac{4e^2n_0}{M_e^*}\mathbf{A}_R, \tag{4.7.2}$$

(where M_e^* is the bipolaron effective mass), which leads to Meissner effect where n_0 is the concentration of superconducting current carriers: $n_0 = N_0/V$. Comparing (4.7.2) with the well-known phenomenological expression for a surface current $\mathbf{j}_S$ (Schmidt, 1997):

$$\mathbf{j}_S = -\frac{c}{4\pi\lambda^2}\mathbf{A}, \tag{4.7.3}$$

and assuming that $\mathbf{A} = \mathbf{A}_R$ from equality $\mathbf{j} = \mathbf{j}_S$ and (4.7.2) and (4.7.3) we derive a well-known expression for the London penetration depth λ:

$$\lambda = \left(\frac{M_e^* c^2}{16\pi e^2 n_0}\right)^{1/2}. \tag{4.7.4}$$

The equality of "microscopic" current expression (4.7.2) to its "macroscopic" value cannot be exact. Accordingly, the equality $\mathbf{A} = \mathbf{A}_R$ is also approximate since $\mathbf{A}_R$ is a vector-potential at the point where the mass center of two electrons occurs, while in the London theory $\mathbf{A}_R$ is a vector potential at the point where the particle occurs. Therefore, it would be more realistic to believe that these quantities are proportional. In this case the expression for the penetration depth will be

$$\lambda = const\left(\frac{M_e^* c^2}{16\pi e^2 n_0}\right)^{1/2} \tag{4.7.4'}$$

where the constant multiplier (of the order of 1 in (4.7.4′)) should be determined from comparison with the experiment.

Expression (4.7.1) was obtained in the case of an isotropic effective mass of current carriers. Actually, it has a more general character and does not change when anisotropy of effective masses is taken into account. For example, in layered HTSC materials the kinetic energy of current carriers in Hamiltonian (4.2.1) should be replaced by the expression:

$$\begin{aligned} T_a = {} & \frac{1}{2m_{||}}\left(\hat{\mathbf{p}}_{1||} - \frac{e}{c}\mathbf{A}_1\right)^2 + \frac{1}{2m_{||}}\left(\hat{\mathbf{p}}_{2||} - \frac{e}{c}\mathbf{A}_2\right)^2 \\ & + \frac{1}{2m_{\perp}}\left(\hat{\mathbf{p}}_{1\perp} - \frac{e}{c}\mathbf{A}_{1z}\right)^2 + \frac{1}{2m_{\perp}}\left(\hat{\mathbf{p}}_{2\perp} - \frac{e}{c}\mathbf{A}_{2z}\right)^2, \end{aligned} \tag{4.7.5}$$

where $\hat{\mathbf{p}}_{1,2||}, \mathbf{A}_{1,2||}$ are the operators of the momentum and vector potential in the planes of the layers (*ab*-planes); $\hat{\mathbf{p}}_{1,2\perp}, \mathbf{A}_{1,2\perp}$ are relevant the values in the direction perpendicular to the planes ((along the c-axis); $m_{||}, m_{\perp}$ are effective masses in the planes and in the perpendicular direction.

As a result of the transformation:

$$\begin{aligned} &\tilde{x} = x, \quad \tilde{y} = y, \quad \tilde{z} = \gamma z, \\ &\tilde{A}_{\tilde{x}} = A_x, \quad \tilde{A}_{\tilde{y}} = A_y, \quad \tilde{A}_{\tilde{z}} = \gamma^{-1} A_z, \\ &\tilde{p}_{\tilde{x}} = p_x, \quad \tilde{p}_{\tilde{y}} = p_y, \quad \tilde{p}_{\tilde{z}} = \gamma^{-1} p_z, \end{aligned} \tag{4.7.6}$$

where $\gamma^2 = m_{\perp}/m_{||}$, γ is an anisotropy parameter, the kinetic energy $\tilde{T}_a$ turns out to be isotropic. Hence, $\tilde{\mathbf{P}}_R + (2e/c)\,\tilde{\mathbf{A}}_{\tilde{R}} = 0$. Then, it follows from (4.7.6) that relation (4.7.1) is valid in an anisotropic case too. It follows that:

$$\begin{aligned} &\mathbf{P}_{R||} = -\frac{2e}{c}\mathbf{A}_{R||}, \qquad \mathbf{P}_{R\perp} = -\frac{2e}{c}\mathbf{A}_{R\perp}, \\ &\mathbf{j}_{||} = 2en_0\mathbf{P}_{R||}/M_{e||}^*, \quad \mathbf{j}_{\perp} = 2en_0\mathbf{P}_{R\perp}/M_{e\perp}^*. \end{aligned} \tag{4.7.7}$$

A magnetic field directed perpendicularly to the plane of the layers will induce currents flowing in the plane of the layers. When penetrating into the sample, this field will attenuate along the plane of the layers. For the magnetic field perpendicular to the plane of the layers $(H_{\perp})$ we denote the London penetration depth by $\lambda_{\perp}$, and for the magnetic field parallel to the plane of the layers $(H_{||})$ by $\lambda_{||}$.

This implies the expression for the London depth of magnetic field penetration into a sample:

$$\lambda_\perp = \left(\frac{M^*_{e\perp}c^2}{16\pi e^2 n_0}\right)^{1/2}, \qquad \lambda_{||} = \left(\frac{M^*_{e||}c^2}{16\pi e^2 n_0}\right)^{1/2}. \tag{4.7.8}$$

For $\lambda_{||}$ and $\lambda_\perp$ the denotations λ_{ab} and λ_c are often used. It follows from (4.7.8) that:

$$\frac{\lambda_\perp}{\lambda_{||}} = \left(\frac{M^*_{e\perp}}{M^*_{e||}}\right)^{1/2} = \gamma^* \tag{4.7.9}$$

It also follows from (4.7.8) that the London penetration depth depends on the temperature:

$$\lambda^2(0)/\lambda^2(T) = n_0(T)/n_0(0). \tag{4.7.10}$$

In particular, for $\omega = 0$ with the use of (4.5.4) we get $\lambda(T) = \lambda(0)\left(1 - (T/T_c)^{3/2}\right)^{-1/2}$. This dependence is compared with other approaches in Section 5.1.

It is usually believed that a Bose system becomes superconducting because of an interaction between the particles. The occurrence of a gap in the spectrum of TI bipolarons can lead to their condensation even when the particles do not interact and the Landau superfluidity criterion:

$$v = \hbar\omega_k/P \tag{4.7.11}$$

(where P is a specific momentum of the bipolaron condensate) can be fulfilled even for noninteracting particles. From condition (4.7.11) we can derive the expression for the maximum value of the current density $j_{\max} = env_{\max}$:

$$j_{\max} = en_0\sqrt{\frac{2\hbar\omega_0}{M_e^*}}. \tag{4.7.12}$$

In conclusion it should be noted that all of the aforesaid refers to local electrodynamics. Accordingly, expressions obtained for λ are valid only on condition that $\lambda \gg \xi$, where ξ is a correlation length which determines the characteristic size of a pair, that is, the characteristic scale of changes in the wave function $\psi(r)$ in (4.2.9). As a rule, this condition is fulfilled in HTSC. In ordinary superconductors the reverse inequality is fulfilled. A nonlocal generalization of superconductor electrodynamics was performed by Pippard (1950). It implies that the relation between $\mathbf{j}_S$ and $\mathbf{A}$ in expression (4.7.3) can be written in the form:

$$\mathbf{j}_S = \int Q(r - r')\mathbf{A}(r')dr', \tag{4.7.13}$$

where is a certain operator whose radius of action is usually believed to be equal to ξ. In the limit $\xi \gg \lambda$ this leads to an increase in the absolute value of the depth of the magnetic field penetration into a superconductor which becomes equal to $(\lambda^2\xi)^{1/3}$ (Pippard, 1950).

4.8 Thermodynamic properties of a TI bipolaron gas in a magnetic field

The fact that Bose condensation of an ideal Bose gas in a magnetic field is impossible (Schafroth, 1955) does not mean that BEC mechanism cannot be used to describe superconductivity in a magnetic field. This follows from the fact that a magnetic field in a superconductor is identically zero. At the same time, abstracting ourselves from SC problem, there are no obstacles to consider a Bose gas to be placed in a magnetic field. Of interest is to investigate this problem with respect to a TI bipolaron gas.

First, it should be noted, that from expression for $\tilde{\omega}_H$, given by (4.5.3), it follows that for $\omega_0 = 0$ Bose condensation of TI bipolarons turns out to be impossible if $H \neq 0$. For an ordinary ideal charged Bose gas, this conclusion was first made in Schafroth (1955). In view of the fact that in the spectrum of TI bipolarons there is a gap between the ground and excited states (Section 4.5), for a TI bipolaron gas this conclusion is invalid at $\omega_0 \neq 0$.

From the expression for $\tilde{\omega}_H$ in (4.5.3) it follows that there is a maximum value of the magnetic field $H_{\max}$ equal to

$$H_{\max}^2 = \frac{2\omega_0\hbar^2 M_e}{\eta^2}. \tag{4.8.1}$$

For $H > H_{\max}$ a homogeneous superconducting state is impossible. As suggested by (4.2.11) and (4.2.16), the quantity η consists from two parts $\eta = \eta' + \eta''$. The value of η' is determined by the integral involved into the expression for $\tilde{\omega}_k$ (4.2.16). Therefore η' depends on the shape of a sample surface. The value of η'' is determined by the sum involved into the expression for $\tilde{\omega}_k$ (4.2.16), and depends on the shape of a sample surface only slightly. Hence, the value of η can change as the shape of a sample surface changes thus leading to a change in $H_{\max}$. With the use of (4.8.1) $\tilde{\omega}_H$ determined by (4.5.3) will be written as

$$\tilde{\omega}_H = \tilde{\omega}\left(1 - H^2/H_{\max}^2\right). \tag{4.8.2}$$

For a given temperature T let us write $H_{cr}(T)$ for the value of the magnetic field for which the superconductivity disappears. This value of the field, according to (4.8.2), corresponds to $\tilde{\omega}_{H_{cr}}$:

$$\tilde{\omega}_{H_{cr}}(T) = \tilde{\omega}\left(1 - H_{cr}^2(T)/H_{\max}^2\right). \tag{4.8.3}$$

The temperature dependence of $\tilde{\omega}_{H_{cr}}(T)$ can be found from condition (4.5.3):

$$C_{bp} = \tilde{T}^{3/2} F_{3/2}\left(\tilde{\omega}_{H_{cr}}(\tilde{T})/\tilde{T}\right).$$

It has the shape shown in Fig. 4.2.

With the use of (4.7.1) and the temperature dependence given in Fig. 4.2 we can find the temperature dependence of $H_{cr}(\tilde{T})$:

$$\frac{H_{cr}^2(\tilde{T})}{H_{\max}^2} = 1 - \frac{\omega_{H_{cr}}(\tilde{T})}{\tilde{\omega}}. \tag{4.8.4}$$

For $\tilde{T} \le \tilde{T}_{ci}$ these dependencies are given in Fig. 4.10.

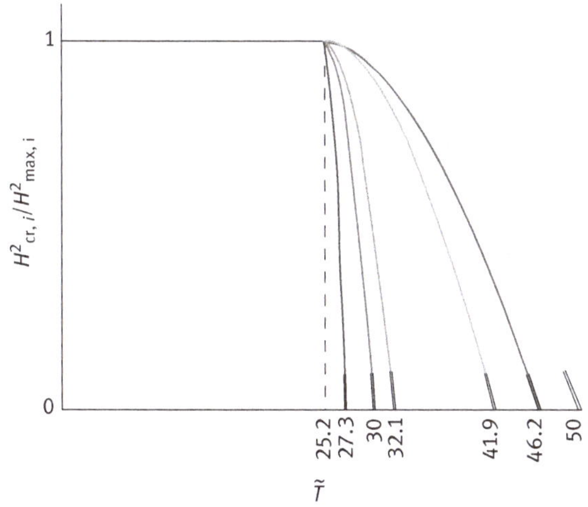

Fig. 4.10: Temperature dependence $H_{cr,i}^2/H_{\max,i}^2$ on the interval $[0; T_{c,i}]$ for the parameter values of $\tilde{\omega}_i$, given in Fig. 4.2.

According to Fig. 4.10, $H_{cr}(\tilde{T})$ reaches its maximum at a finite temperature $\tilde{T}_c(\tilde{\omega}=0) \le \tilde{T}_c(\omega_{0i})$. Figure 4.10 suggests that at a temperature below $\tilde{T}_c(\tilde{\omega}=0) = 25.24$ a further decrease of the temperature no longer changes the critical field $H_{cr}(\tilde{T})$ irrespective of the gap value $\tilde{\omega}$.

Let us introduce a concept of a transition temperature $T_c(H)$ in a magnetic field H. Figure 4.11 shows the dependencies $T_c(H)$ resulting from Fig. 4.10 and determined by the relations:

$$C_{bp} = \tilde{T}_{c,i}^{3/2}(H) F_{3/2}\left(\tilde{\omega}_{H,i}/\tilde{T}_{c,i}(H)\right), \qquad \tilde{\omega}_{H,i} = \tilde{\omega}_{H=0,i}\left[1 - H^2/H_{\max,i}^2\right].$$

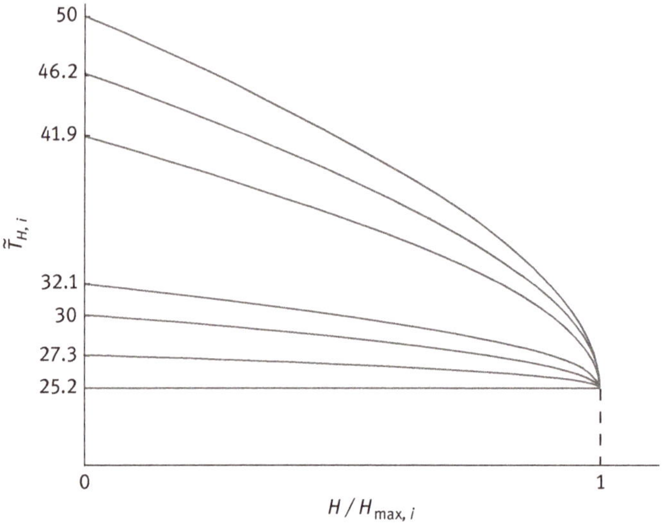

Fig. 4.11: Dependence of the critical transition temperature $\tilde{T}_{H,i}$ on the magnetic field H for the parameter values of $\tilde{\omega}_i$ given in Fig. 4.2.

Figure 4.11 suggests that the critical temperature of the transition $\tilde{T}_c(H)$ changes in a stepwise fashion as the magnetic field reaches the value $H_{\max,i}$.

To solve the problem of the type of a phase transition in a magnetic field we will proceed from the well-known expression which relates the free energies in the superconducting and normal states:

$$F_S + H^2/8\pi = F_N, \tag{4.8.5}$$

where F_s and F_N are free energies of the unit volume of superconducting and normal states, respectively:

$$F_S = \frac{N}{V}E_{bp}(H=0) - \frac{2}{3}\Delta E(\omega_{H=0})\frac{N}{V}, \qquad F_N = \frac{N}{V}E_{bp}(H) - \frac{2}{3}\Delta E(\omega_H)\frac{N}{V},$$

where $\Delta E = E - E_{bp}$, $E = \omega^* \tilde{E}$, where $\tilde{E}$ is determined by the formula (4.5.8). Differentiating (4.8.5) with respect to temperature and taking into account that $S = -\partial F/\partial T$ we express the transition heat q as

$$q = T(S_N - S_S) = -T\partial(F_N - F_S)/\partial T = -T\frac{H_{cr}}{4\pi}\frac{\partial H_{cr}}{\partial T}. \tag{4.8.6}$$

Accordingly, the entropy difference $S_S - S_N$ is

$$S_S - S_N = \frac{H_{cr}}{4\pi}\left(\frac{\partial H_{cr}}{\partial T}\right) = \frac{H^2_{\max}}{8\pi\omega^*}\left(\tilde{S}_S - \tilde{S}_N\right). \tag{4.8.7}$$

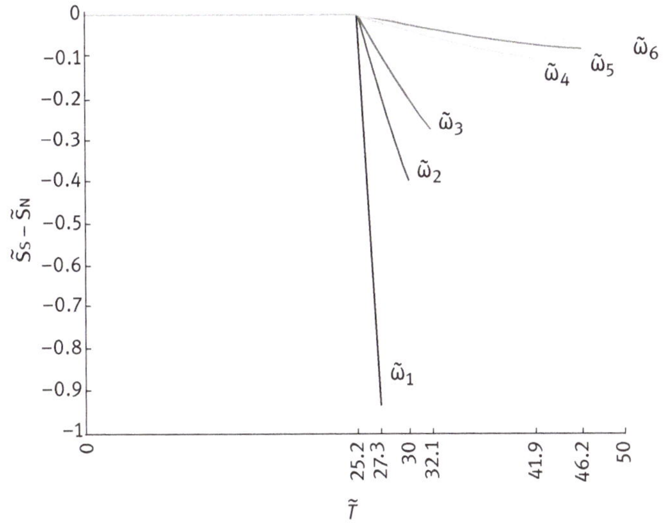

Fig. 4.12: Temperature dependencies of differences between the entropies in superconducting and normal states for $\tilde{\omega}_i$ parameters used in Figs. 4.10 and 4.11.

Figure 4.12 illustrates the temperature dependence of differences between the entropies in the superconducting and normal states (4.8.7) for different values of critical temperatures $T_c(\tilde{\omega}_i)$, given in Fig. 4.2. The differences presented in Fig. 4.12 can seem strange in, at least, two respects:

1. In the BCS and Landau theories at the critical point T_c the entropy difference vanishes according to Rutger's formula. Entropy in Fig. 4.12 is a monotonous function $\tilde{T}$ and does not vanish at $T = T_c$.
2. The absolute value of the difference $|\tilde{S}_S - \tilde{S}_N|$ when approaching the limit point which corresponds to $\tilde{\omega} = 0$, would seem to increase rather than to decrease vanishing for $\tilde{\omega} = 0$.

Regarding point 2, this is indeed the case for $|\tilde{S}_S - \tilde{S}_N|$, since the value of the maximum field $H_{\max}$ and, accordingly, the multiplier $H_{\max}/8\pi$ which relates the quantities $S_S - S_N$ and $\tilde{S}_S - \tilde{S}_N$ vanishes for $\tilde{\omega} = 0$.

Regarding point 1, as it will be shown below, Rutger's formula turns out to be inapplicable to a TI bipolaron Bose condensate.

Table 4.3 lists the values of $\tilde{S}_S - \tilde{S}_N$ for critical temperatures corresponding to different values of $\tilde{\omega}_{H_{cr,i}}$.

The results obtained lead to some fundamental consequences:

1. The curve of $H_{cr}(T)$ dependence (Fig. 4.10) has a zero derivative, $dH_{cr}(T)/dT = 0$ for $T = 0$. This result is in accordance with the Nernst theorem which implies that entropy determined by (4.8.6) is equal to zero at $T = 0$.

2. According to Fig. 4.10, $H_{cr}(T)$ monotonously decreases as T grows for $T > T_c(\tilde{\omega} = 0)$, and does not change for $T \leq T_c(\tilde{\omega} = 0)$. Hence $\partial H_{cr}(T)/\partial T < 0$ for $T > T_c(\tilde{\omega} = 0)$. Therefore, on the temperature interval $[T_c(\tilde{\omega} = 0), T_c(\tilde{\omega})]$ $S_S < S_N$, and on the interval $[0, T_c(\tilde{\omega} = 0)]$ $S_S = S_N$.

This has some important implications:
1. Transition on the interval $[0, T_c(\tilde{\omega} = 0)]$ occurs without absorption or liberation of latent heat since in this case $S_S = S_N$. That is, in the experiment, it will be seen as a second-order phase transition. Actually, on the interval $[0, T_c(\tilde{\omega} = 0)]$, the phase transition into the superconducting state is an infinite-order phase transition since in this region any-order derivatives of the energy difference $F_S - F_N$, vanish according to (4.8.5) and Fig. 4.10.
2. In a magnetic field on the interval $[T_c(\tilde{\omega} = 0), T_c(\tilde{\omega})]$, which corresponds to $S_S < S_N$, a transition from a superconducting to a normal state occurs with absorption of latent heat. On the contrary, in passing on from a normal to a superconducting state latent heat releases. The phase transition on the interval $[0, T_c(\tilde{\omega} = 0)]$ is not attended by a release or absorption of latent heat, being an infinite-order phase transition.

With regard to the fact that the heat capacity of a substance is determined by the formula $C = T(\partial S/\partial T)$, a difference in the specific heat capacities in a superconducting and normal states, according to (4.8.7), is written as

$$C_S - C_N = \frac{T}{4\pi}\left[\left(\frac{\partial H_{cr}}{\partial T}\right)^2 + H_{cr}\frac{\partial^2 H_{cr}}{\partial T^2}\right]. \qquad (4.8.8)$$

The well-known Rutger's formula can be obtained from this expression if we put in (4.8.8) the critical field $H_{cr}(T_c) = 0$ for $T = T_c$and leave in the bracket in the right-hand side of (4.8.8) only the first term:

$$(C_S - C_N)_R = \frac{T}{4\pi}\left(\frac{\partial H_{cr}}{\partial T}\right)^2_{T_c}.$$

It is easy to see that at the point $T = T_c$ the value of $\omega_{H_{cr}}$, determined by Fig. 4.2, for all the temperature values has a finite derivative with respect to T and therefore, according to (4.8.4), an infinite derivative $\partial H_{cr}/\partial T$ for $T = T_c$. Hence, the second term in the square bracket in (4.8.8) turns to $-\infty$, leaving this bracket a finite value. As a result, the difference in the heat capacities in our model of Bose gas is properly expressed as

$$C_S - C_N = \frac{T}{8\pi}\frac{\partial^2}{\partial T^2} H_{cr}^2(T) = \frac{H_{\max}^2}{8\pi\omega*}(\tilde{C}_S - \tilde{C}_N),$$

$$\tilde{C}_S - \tilde{C}_N = \tilde{T}\frac{\partial^2}{\partial \tilde{T}^2}(H_{cr}^2(T)/H_{\max}^2). \quad (4.8.9)$$

Table 4.3 lists the values of the quantity $\tilde{C}_S - \tilde{C}_N$ for the values of the critical temperatures corresponding to different values of $\tilde{\omega}_{H_{cr,i}}$. It should be noted that according to the results obtained, the maximum of the heat capacity jump occurs at a zero magnetic field and decreases as the magnetic field grows, vanishing for $H = H_{cr}$, which is fully consistent with the experimental data (Section 5.1). Comparison of the heat capacity jumps shown in Fig. 4.5, with expression (4.8.9) enables us to calculate the value of $H_{\max}$. The values of $H_{\max}$ obtained for different values of ω_i are listed in Tab. 4.3. These values unambiguously determine the values of constant η in formulae (4.3.7′) and (4.3.7″).

Tab. 4.3: The values of $H_{\max}$, entropy differences $\tilde{S}_S - \tilde{S}_N$, and heat capacities $\tilde{C}_S - \tilde{C}_N$ in a superconducting and normal states determined by relations (4.8.7) and (4.8.9) are presented for the transition temperatures $\tilde{T}_{C_i}$ for the same values of $\tilde{\omega}_{H_{cr,i}}$ as in Fig. 4.2.

i	$\tilde{\omega}_{H_{cr,i}}$	$\tilde{T}_{C_i}$	$\tilde{S}_S - \tilde{S}_N$	$\tilde{C}_S - \tilde{C}_N$	$H_{\max} \times 10^{-3}$ Oe
0	0	25.2	0	0	0
1	0.2	27.3	−0.94	−11.54	2.27
2	1	30	−0.4	−2.18	7.8
3	2	32.1	−0.27	−1.05	13.3
4	10	41.9	−0.1	−0.19	47.1
5	15	46.2	−0.08	−0.12	64.9
6	20	50	−0.07	−0.09	81.5

It follows from what has been said that Ginzburg–Landau temperature expansion for a critical field near a critical temperature T_c is inapplicable to a TI bipolaron Bose condensate. Since the temperature dependence $H_{cr}(T)$ determines the temperature dependencies of all the thermodynamic quantities, this conclusion is valid for all such quantities. As noted in the Introduction, this conclusion stems from the fact that the BCS theory, being nonanalytical with respect to the coupling constant under no conditions passes on the bipolaron condensate theory.

Above we dealt with the isotropic case. In the anisotropic case it follows from formulae (4.3.7′) and (4.3.7″) that:

$$H_{\max}^2 = H_{\max\perp}^2 = \frac{2\omega_0 M_\perp \hbar^2}{\eta^2}, \quad \mathbf{B}||\mathbf{c}, \tag{4.8.10}$$

– that is, in the case when a magnetic field is directed perpendicularly to the plane of the layers and:

$$H_{\max}^2 = H_{\max||}^2 = \frac{2\omega_0 M_{||} \hbar^2}{\eta^2}, \quad \mathbf{B}\perp\mathbf{c}, \tag{4.8.11}$$

– in the case when a magnetic field lies in the plane of the layers. From (4.8.10) and (4.8.11), it follows that:

$$\frac{H_{\max\perp}^2}{H_{\max||}^2} = \sqrt{\frac{M_\perp}{M_{||}}} = \gamma. \tag{4.8.12}$$

With the use of (4.8.4), (4.8.11), and (4.8.12), we get for the critical field $H_{cr}(\tilde{T})$ in the directions perpendicular and parallel to the plane of the layers:

$$H_{cr||,\perp}(\tilde{T}) = H_{\max||,\perp}\sqrt{1 - \omega_{H_{cr}}(\tilde{T})/\tilde{\omega}}. \tag{4.8.13}$$

It follows from (4.8.13) that relations $H_{cr||}(\tilde{T})/H_{cr\perp}(\tilde{T})$ are independent of temperature. The dependencies obtained are compared with the experiment in Chapter 5.

4.9 Translation-invariant bipolarons and a pseudogap phase

Among the most amazing and mysterious phenomena of high-temperature superconductivity (HTSC) is the existence of a pseudogap phase at a temperature above the critical temperature of a superconducting (SC) transition (Norman et al., 2005; Vishik et al., 2010; Timusk and Statt, 1999; Huefner et al., 2008; Lee et al., 2006). In a pseudogap phase, the spectral density of states near the Fermi surface demonstrates a gap for $T > T_C$, where T_c is a temperature of a SC transition which persists up to the temperatures T^* $(T^* > T_C)$, above which the pseudogap disappears. Presently, the explanation of this phenomenon is reduced to two possibilities. According to the first one, it is believed that for $T > T_C$ some incoherent electron pairs persist in the sample, while for $T < T_c$ their motion becomes coherent and they pass on to the SC state. For $T > T^*$ the pairs disintegrate and the pseudogap state disappears (Randeria and Trivedi, 1998; Franz, 2007; Emery and Kivelson, 1995; Curty and Beck, 2003). According to the second one, the transition to the pseudogap phase is not concerned with superconductivity, but is caused by the formation of a certain phase with a hidden order parameter or a phase with spin fluctuations (Moon and Sachdev, 2009; Sadovskii, 2001; Bardeen et al., 1957).

Presently the first viewpoint on the nature of a pseudogap in HTSC increasingly dominates which is associated with the idea that paired electron states exist for $T > T_C$. The question of the nature of paired electron states per se remains open. In this book paired electron states are taken to be TI bipolarons.

The TI bipolaron theory of SC based on the Pekar–Fröhlich Hamiltonian of EPI when EPI cannot be considered to be weak as distinct from the Bardeen–Cooper–Schrieffer theory (Bardeen et al., 1957) was developed in Lakhno (2018, 2019a, 2019b) (see also review of Lakhno, 2020a). The role of Cooper pairs in this theory belongs to TI bipolarons whose size (≈ 1nm) is much less than that of Cooper pairs ($\approx 10^3$nm). According to this theory in HTSC materials TI bipolarons are formed near the Fermi surface and represent a charged Bose gas capable of experiencing BEC at high critical temperature which determines the temperature of a SC transition.

As distinct from Cooper pairs, TI bipolarons have their own excitation spectrum:

$$E_k^{bp} = E_{bp}\Delta_{k,0} + \left(\omega_0 + E_{bp} + k^2/2M_e\right)\left(1 - \Delta_{k,0}\right), \tag{4.9.1}$$

$$M_e = 2m, \Delta_{k,0} = 1 \text{ for } k = 0 \text{ and } \Delta_{k,0} = 0 \text{ for } k \neq 0$$

where E_{bp} is the ground state energy of a TI bipolaron (reckoned from the Fermi level), ω_0 is the frequency of an optical phonon, m is a mass of a band electron (hole), and k is a wave vector numbering excited states of a TI bipolaron.

This spectrum has a gap which in the isotropic case is equal to the frequency of an optical phonon ω_0. At that, the inequality $\omega_0 \gg |E_{bp}|$ corresponds to the case of a weak EPI, $\omega_0 \ll |E_{bp}|$ – to the case of strong coupling and $\omega_0 \sim |E_{bp}|$ – to the case of intermediate coupling. According to Lakhno (2018, 2019a, 2019b, 2020a) the number of TI bipolarons N_{bp} at temperature $T = 0$ is equal to $N_{bp} \cong N\omega_0/2E_F$, where N is the total number of electrons (holes), E_F is the Fermi energy, that is $N_{bp} \ll N$.

The scenario of a SC based on the idea of a TI bipolaron as a fundamental boson responsible for superconducting properties explains many thermodynamic and spectroscopic properties of HTSC. For this reason, the problem of the temperature of a transition T^* to the pseudogap state is of interest.

4.9.1 Temperature of a pseudogap phase

Obviously, the temperature of a transition from a pseudogap phase to a normal one T^* in this model is determined by disintegration of TI bipolarons into individual TI polarons. Thermodynamically, the value of T^* should be determined from the condition that the free energy of a TI bipolaron gas exceeds the free energy of a TI polaron gas determined by the spectrum of TI polarons:

$$E_k^P = E_p \Delta_{k,0} + \left(\omega_0 + E_p + k^2/2m\right)\left(1 - \Delta_{k,0}\right) \tag{4.9.2}$$

where E_p is the energy of the polaron ground state.

For further consideration, it is significant that the number of TI bipolarons in HTSC compounds is $N_{bp} \ll N$. For $n = N/V = 10^{21}\text{cm}^{-3}$, where V is the system volume, the typical values of n_{bp} are of the order of $n_{bp} \sim 10^{18} - 10^{19}\text{cm}^{-3}$. Taking into account that $T^* > T_C$, in order to calculate the statistical sum of the bipolaron gas Z_{bp} in the vicinity of T^* one can use a classical approximation which requires that in the region of stability of the bipolaron gas the inequality:

$$T^* > T > T_C \tag{4.9.3}$$

be fulfilled. In this case the expression for the statistical sum of the TI bipolaron gas has the form:

$$Z_{bp} = \frac{1}{h^{3N_{bp}} N_{bp}!} \prod_{i=1}^{N_{bp}} \int d^3k_i e^{-E_{k_i}^{bp}/T} = \left[e^{-(\omega_0 + E_{bp})/T} \left(\frac{2\pi M_e T}{h^2}\right)^{3/2} \frac{eV}{N_{bp}}\right]^{N_{bp}} \tag{4.9.4}$$

where $e \approx 2.781$ is the natural logarithm base, $h = 2\pi\hbar$ is Planck constant.

Accordingly, for the statistical sum of a TI polaron gas formed as a result of disintegration of TI bipolarons, similarly to (4.9.4), we get

$$Z_p = \left[e^{-(\omega_0 + E_p)/T} \frac{(2\pi m T)^{3/2}}{h^3} \frac{eV}{2N_{bp}}\right]^{2N_{bp}} \tag{4.9.5}$$

The condition of stability of a TI bipolaron gas with respect to its decay into a TI polaron gas is written as

$$Z_{bp} \geq Z_p \tag{4.9.6}$$

where the equality describes the case of an equilibrium between the two gases which corresponds to the equation for temperature T^* of a transition from a normal phase to a pseudogap one.

Substitution of (4.9.4) and (4.9.5) into (4.9.6) leads to the condition:

$$\Delta = |E_{bp}| + \omega_0 - 2|E_p| \geq \frac{3}{2} T \ln \kappa T, \qquad \kappa = \left(\frac{e}{4}\right)^{2/3} \frac{\pi m}{n_{bp}^{2/3} h^2} \tag{4.9.7}$$

In the case of equality expression (4.9.7) yields the equation for determining T^*:

$$z = W e^W, \qquad T^* = \kappa^{-1} e^W, \qquad z = 2\kappa\Delta/3. \tag{4.9.8}$$

Fig. 4.13 shows the solution $W = W(z)$ (Lambert function) (4.9.8) on condition that limitation (4.9.3) is fulfilled.

It holds on the interval $-e^{-1} < z < \infty$. On the interval $-e^{-1} < z < 0$ Lambert function is negative. Requirement (4.9.3) leads to the condition: $-2{,}28 < W < \infty$. Taking into account the expression for the temperature of a SC transition obtained in Section 4.5:

$$T_C = T_C(\omega_0) = \left(F_{3/2}(0)/F_{3/2}(\omega_0/T_C)\right)^{3/2} T_C(0), \tag{4.9.9}$$

$$T_C(0) = 3.31\hbar^2 n_{bp}^{2/3}/M_e, \qquad F_{3/2}(x) = \frac{2}{\sqrt{\pi}}\int_0^\infty \frac{t^{1/2}dt}{e^{t+x}-1},$$

we express T^* from (4.9.8) as

$$T^* \approx 9.8\left(F_{3/2}(\omega_0/T_C)/F_{3/2}(0)\right)^{2/3} T_C \exp W. \tag{4.9.10}$$

Thus for example, for $\omega_0 \approx T_C$ we obtain from (4.9.10) that $T^* \approx 3T_C(1)\exp W$, where $T_C(1)$ is determined by (4.9.9): $T_C(1) \approx 3.3T_C(0)$, that is for W=0 the pseudogap temperature T^* exceeds the temperature of a SC transition T_C more than threefold.

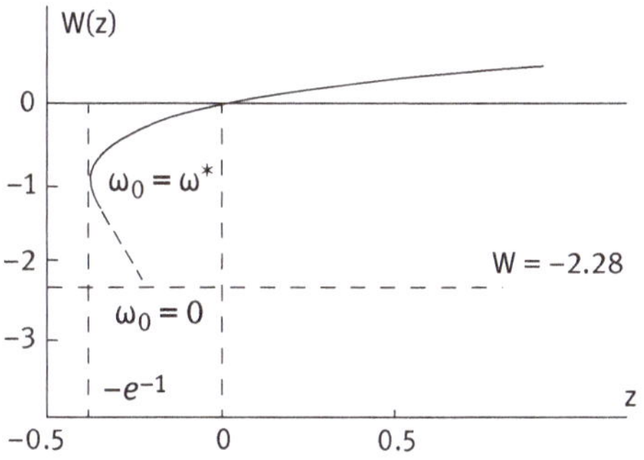

Fig. 4.13: Solution $W(z)$ of equation (4.9.8).

For $\omega_0 \gg T_C$ the temperature of a pseudogap phase is $T^* \gg T(1)$. In this case for estimating T^* we can use an approximate formula derived from (4.9.8):

$$T^* \approx \frac{2}{3}\Delta/\ln\frac{2}{3}\kappa\Delta, \qquad \kappa|\Delta| > 3/2. \tag{4.9.11}$$

This limit, however, is observed in experiments only rarely. It can be concluded that in HTSC materials the main contribution into EPI leading to SC is made by phonon frequencies with $\omega_0 < T_C(10\text{ meV})$. This estimate is an order of magnitude less than the estimates of phonon frequencies which are generally believed to make the main contribution into the SC. It also follows from (4.9.11) that T^* grows only logarithmically as the concentration of n_{bp} increases, while $T_C \tilde{n}_{bp}^{2/3}$. Hence for a certain value of n_{bp} the condition $T^* < T_C$, can be fulfilled which corresponds to disappearance of a pseudogap phase as it follows from the form of the exact solution of equation (4.9.8) (see nevertheless Chapter 7). In HTSC materials this takes place as doping increases to an optimal value for which the pseudogap phase no longer exists.

4.9.2 Isotope coefficient for the pseudogap phase

The TI bipolaron theory of the pseudogap phase developed above enables one to investigate its isotopic properties.

It follows from (4.9.8) that like the SC phase, the pseudogap one possesses the isotopic effect. According to (4.9.8), the isotope coefficient is:

$$\alpha^* = -d\ln T^* / d\ln M \tag{4.9.12}$$

where M is the mass of an atom replaced by its isotope. With regard to the fact that $\omega_0 \sim M^{-1/2}$, it takes the form:

$$\alpha^* = \frac{\omega_0}{3T^*} \frac{1}{1+W(z)}, \tag{4.9.13}$$

It follows from (9.13) and Fig. 4.13 that for the lower branch: $W(z) < -1$ and $\alpha^* < 0$. Accordingly, for the upper branch: $W(z) > -1$ and $\alpha^* > 0$. It should be noted that the upper branch corresponds to $\omega_0 > \omega^*$, while the lower branch to $\omega_0 < \omega^*$, where $\omega^* \approx T_C$. Expressions (4.9.10) and (4.9.13) yield:

$$W = \ln\left[c\left(F_{3/2}(0)/F_{3/2}(\omega_0/T_C)\right)^{2/3} T^* / T_C\right], \tag{4.9.14}$$

where $c \approx 0.1$ Fig. 4.14 illustrates the graph of the dependence $\alpha^*(\omega_0)$.

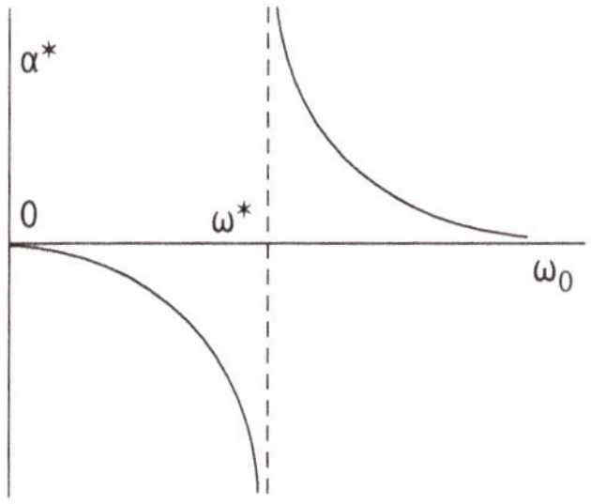

Fig. 4.14: Dependence of the isotope coefficient for the pseudogap temperature T^* on the phonon frequency $\omega_0(\omega^* \approx T_c)$.

Hence, depending on the value of the phonon frequency ω_0 coefficient α^* can have any sign and the value: $\alpha^* < 0$ for $\omega_0 < \omega^*$ and $\alpha^* > 0$ for $\omega_0 > \omega^*$. For $\omega_0 = \omega^*$, the isotope coefficient becomes infinite: $\alpha^*(\omega^* \pm 0) = \pm\infty$. The results obtained suggest that the isotope exponent diverges for $\omega_0 \to \omega^*$, that is for $T^* \approx T_c$ (Fig. 4.14). Great negative values of the isotope coefficient in the pseudogap state were observed experimentally in Rubio Temprano et al. (2000), Bendele et al. (2017), and Furrer (2005). It should be noted that in some cases negative values of the isotope coefficient were also observed in ordinary SC (Bill et al., 1998a, 1998b), which exceeded in modulus the value of the isotope coefficient in monoatomic systems $\alpha = 0{,}5$, yielded by the BCS. According to the theory suggested this is possible for $T^* \approx T_c$.

4.9.3 Isotope coefficient for pseudogap phase in magnetic field

According to (4.3.6), the spectrum of a TI bipolaron in a magnetic field **B** is determined by the modified expression (4.9.1):

$$E_k^{bp} = E_{bp}\Delta_{k,0} + \left(\omega_0 + E_{bp} + k/2M_e + \eta_{bp}(\mathbf{Bk})/2M_e\right)(1 - \Delta_{k,0}), \tag{4.9.15}$$

where the quantity η_{bp}, according to (4.8.1), is related to the first critical field SC $B_{\max}$ as

$$\eta_{bp} = \sqrt{\frac{2\omega_0 M_e}{B_{\max}}}.$$

The spectrum of a TI polaron will be determined by relation (4.9.2), since in weak fields $B < B_{\max}$ the magnetic field leaves it practically unchanged ($B_{\max}$, being the first critical field, is always much less than the value of the quantizing magnetic field of a TI polaron).

Performing calculations similar to Section 4.2 we obtain the same relations as in Sections 4.2 and 4.3, where ω_0 should be replaced by $\tilde{\omega}_0 = \omega_0\left(1 + B^2/B_{\max}^2\right)$. Thus, for example, the graph of the dependence $\alpha(\tilde{\omega}_0)$, determined by Fig. 4.14, in a magnetic field will be a graph of the dependence $\alpha(\tilde{\omega}_0)$. Hence, for $\tilde{\omega}_0 < \omega^*$ an increase in the field value will lead to larger negative values of the isotope coefficient, and for $\tilde{\omega}_0 > \omega^*$ to smaller positive values of α^*. In particular, a situation is possible when, for a certain value of the field, the isotope coefficient, being negative, as the field increases, becomes infinite at the point $\tilde{\omega}_0 = \omega^*$ and becomes positive for $\tilde{\omega}_0 > \omega^*$. It will be interesting to verify these conclusions experimentally.

4.9.4 Discussion

Here, we have shown that the existence of the pseudogap state and nonstandard behavior of the isotope coefficient in HTSC materials can be explained on the basis of the EPI without the involvement of any other scenarios (Labbé and Bok, 1987; Radtke and Norman, 1994; Schüttler and Pao, 1995; Nazarenko and Dagotto, 1996; Greco and Zeyner, 1999).

We witness further discussion on the nature of the pseudogap phase in HTSC materials. It follows from the foregoing consideration that the pseudogap is a universal effect and should arise as TI bipolarons are formed in a system. The fact that for a long time the occurrence of the pseudogap phase was associated with side effects is caused by that this phase is observed even in ordinary SC (Medicherla et al., 2007; Chainani et al., 2001; Yokoya et al., 2002; Sacépé et al., 2010; Mondal et al., 2011; Thakur et al., 2013), where the occurrence of the pseudogap was explained by a crystallographic disorder or reduced dimensionality which are usually observed in disordered metals.

In connection to this, recent experiments with MgB_2 HTSC seem to be of importance (Patil et al., 2017). As distinct from oxide ceramics, MgB_2 does not have a magnetic order and, as proponents of an external nature of the pseudogap state suggest, should not have a pseudogap. To exclude any other possibilities concerned with disorder, low-dimensionality effects, and so on in experiment (Patil et al., 2017) use was made of highly perfect crystals. Experiments made in Patil et al. (2017) convincingly demonstrated the availability of the pseudogap state in MgB_2 and responsibility of EPI for this state. The results obtained provide good evidence for the TI bipolaron mechanism of the formation of the pseudogap state.

The simple scenario presented in the paper may overlap with the effects associated with spin fluctuations, formation of charge density waves (CDW) and spin density waves (SDW), pair density waves (PDW) and bond density waves (BDW), formation of stripes (e.g., a giant isotopic effect caused by EPI was observed in $La_{2-x}Sr_xCuO_4$ in the vicinity of the temperature of charged stripe ordering when replacing ^{16}O by ^{18}O (Lanzara et al, 1999)), clusters, other types of interactions, and so on. The suggested TI bipolaron mechanism of the pseudogap phase formation and explanation of isotopic effects in HTSC materials on its basis are also important in view of universality of this mechanism.

At present, there are only a few experiments to study the isotope effect in the pseudogap phase. The experiments on the influence of the magnetic field on the isotope coefficient proposed in the book for the temperature of the transition to the pseudogap phase are new. The agreement of the experimental results with the theoretical predictions would indicate the validity of the assumption of the TI bipolaron mechanism of HTSC.

In Lakhno (2020b) an assumption was made that a CDW, and, accordingly, PDW have a bipolaron nature in HTSC (Chapter 6). If we take this assumption, then the quantity E_{bp} involved in (9.1) and (9.15) should refer to quantity $E_{bp}(P_{CDW})$, where P_{CDW} is the wave vector of the CDW: $P_{CDW} = P_{PDW}$. Some of TI bipolarons which

can exist at a temperature exceeding the temperature of CDW formation have a momentum different from P_{CDW} and therefore a finite life time, determining a "smeared" transition between the pseudogap and normal phases.

From the theory developed it follows that the temperature of SC transition can be increased if we increase the concentration of bipolarons, for example

by terahertz irradiation with the energy of about of the phonon TI polaron energy gap (Mankowsky et al., 2014; Mitrano et al., 2016). Other methods of increasing SC transition temperature will be considered in Chapter 5.

4.10 Scaling relations

Scaling relations play an important role in the superconductivity theory by promoting a search for new high-temperature superconductors with record parameters. Such relations can be a generalization of a lot of experiments having no reliable theoretical justification, or can be deduced from less-than-reliable theoretical construct, though being experimentally confirmed in the future. An example is provided by Uemura law considered in the next section 5.1.

The theory presented enables one to give a natural explanation to some important scaling relations. In particular, in this section we deduce Alexandrov formula (Alexandrov, 1999; Alexandrov and Kabanov, 1999) and Homes' scaling law.

4.10.1 Alexandrov formula

As noted in Section 4.3, in an anisotropic case formula (4.5.3) takes on the form:

$$\tilde{T}_c = F_{3/2}^{-2/3}\left(\tilde{\omega}/\tilde{T}_c\right)\left(\frac{n_{bp}}{M_{||}}\right)^{2/3}\frac{2\pi\hbar^2}{M_{\perp}^{1/3}\omega*}. \tag{4.10.1}$$

It is convenient to pass on from the quantities which are difficult to measure in an experiment n_{bp}, $M_{||}$, $M_{\perp}$ to those which can be measured experimentally:

$$\lambda_{ab} = \left[\frac{M_{||}}{16\pi n_{bp}e^2}\right]^{1/2}, \quad \lambda_c = \left[\frac{M_{\perp}}{16\pi n_{bp}e^2}\right]^{1/2}, \quad R_H = \frac{1}{2en_{bp}}, \tag{4.10.2}$$

where $\lambda_{ab} = \lambda_{||}$, $\lambda_c = \lambda_{\perp}$ are London depths of penetration into the planes of the layers and in the perpendicular direction, accordingly; R_H is Hall constant. In expressions (4.10.2) the light velocity is assumed to be equal to 1: $c = 1$. With the use of relations (4.10.2) and (4.10.1) we get

$$k_B T_c = \frac{2^{1/3}}{8} F_{3/2}^{-2/3}(\tilde{\omega}/\tilde{T}_c) \frac{\hbar^2}{e^2} \left(\frac{eR_H}{\lambda_{ab}^4 \lambda_c^2} \right)^{1/3} . \qquad (4.10.3)$$

Here the value of eR_H is measured in cm^3, λ_{ab}, λ_c – in cm, T_c – in K.

Taking into account that in most HTSC $\tilde{\omega} \approx \tilde{T}_c$ and function $F_{3/2}(\tilde{\omega}/\tilde{T})$ varies in the vicinity of $\tilde{\omega} \approx \tilde{T}_c$, only slightly, with the use of the value $F_{3/2}(1) = 0{,}428$ we derive from (4.10.3) that T_c is equal to

$$T_c \cong 8.7 \left(\frac{eR_H}{\lambda_{ab}^4 \lambda_c^2} \right)^{1/3} . \qquad (4.10.4)$$

Formula (4.10.4) differs from Alexandrov's formula (Alexandrov, 1999; Alexandrov and Kabanov, 1999) only in a numerical coefficient which is equal to 1.64 in Alexandrov (1999 and Alexandrov and Kabanov (1999). As it is shown in Alexandrov (1999) and Alexandrov and Kabanov (1999), formula (4.10.4) properly describes, almost without exception, a relation between the parameters for all known HTSC materials. It follows from (4.10.1) that Uemura's relation (Uemura et al., 1989; Uemura et al., 1991) is a particular case of formula (4.10.4).

In an isotropic case, formulae (4.10.3) and (4.10.4) also yield a well-known law of a linear dependence of T_c on the inverse value of the squared London penetration depth.

4.10.2 Homes' law

Homes' law claims that in the case of superconducting materials scaling relation holds (Homes et al., 2004; Zaanen, 2004):

$$\rho_S = C\sigma_{DC}(T_C)T_C, \qquad (4.10.5)$$

where ρ_S is the density of the superfluid component for $T = 0$, $\sigma_{DC}(T_C)$ is the direct current conductivity for $T = T_c$, C is a constant equal to $\approx 35\ cm^{-2}$ for ordinary superconductors and HTSC for a current running in the plane of the layers.

The quantity ρ_S in (4.10.5) is related to plasma frequency ω_p as $\rho_S = \omega_p^2$ (Erdmenger et al., 2012) ($\omega_p = \sqrt{4\pi n_S e_S^2/m_S^*}$, where n_S is a concentration of superconducting current carriers; m_S^*, e_S are a mass and a charge of superconducting current carriers). With the use of this relation, relation $\sigma_{DC} = e_n^2 n_n \tau / m_n^*$, where n_n is a concentration of current carriers for $T = T_C$, m_n^*, e_n are a mass and charge of current carriers for $T = T_c$ and relation $\tau \sim \hbar/T_C$, where τ is a minimum Planck time of electron scattering at a critical point (Erdmenger et al., 2012), on the assumption that $e_S = e_n$, $m_S = m_n$, we get from (4.10.5):

$$n_S(0) \cong n_n(T_C). \tag{4.10.6}$$

In our scenario of Bose condensation of TI bipolarons, Homes' law in the form of (4.10.6) becomes almost obvious. Indeed, for $T = T_c$ TI bipolarons are stable (they decay at a temperature equal to the pseudogap energy which far exceeds T_c). Their concentration at $T = T_c$ is equal to n_n and, therefore, at $T = T_c$ these bipolarons start forming a condensate whose concentration $n_S(T)$ reaches maximum $n_S(0) = n_n(T_C)$ at $T = 0$ ((i.e., when bipolarons become fully condensed), which corresponds to relation (4.10.6). It should be noted that in the framework of the BCS theory Homes' law cannot be explained.

5 Comparison with experiment

Most experiments on HTSC can conventionally be divided into thermodynamic and spectroscopic ones. In this chapter we consider the main experiments from both the groups which are explained on the basis of the TI bipolaron theory.

5.1 Thermodynamic experiments

Success of the Bardeen–Cooper–Schrieffer (BCS) theory is concerned with successful explanation of some experiments in ordinary metal superconductors where electron–phonon interaction (EPI) is not strong. It is arguable that EPI in high-temperature ceramics SC is rather strong (Meevasana et al., 2006a, 2006b; Mishchenko et al., 2008), and the BCS theory is hardly applicable to them. In this case it may be worthwhile to use the description of HTSC properties on the basis of bipolaron theory. As is known, Eliashberg theory which was developed to describe SC with strong EPI (Eliashberg, 1960) is inapplicable to describe polaron states (Alexandrov, 2003; Alexandrov and Mott, 1994). Let us list some experiments on HTSC which are in agreement with the translation-invariant (TI) bipolaron theory

According to the main currently available SC theories (BCS, RVB, t-J theories (Bardeen et al., 1957; Anderson, 1997; Izyumov, 1997)), at low temperatures all the current carriers should be paired (i.e., the superconducting electron density coincides with the superfluid one). In recent experiments on overdoped SC (Božović et al., 2016) it was shown that this is not the case – only a small portion of current carriers were paired. The analysis of this situation performed in Zaanen (2016) demonstrates that the results obtained in Božović et al. (2016) do not fit in the available theoretical constructions. The TI bipolaron theory of SC presented above gives an answer to the question of paper (Zaanen, 2016) – where most of the electrons in the studied SC disappeared? The answer is that only a small portion of electrons n_{bp}:$n_{bp} \approx n\omega_0/E_F \ll n$ occurring near the Fermi surface are paired and determine the surface properties of HTSC materials.

Actually, however, the theory of EPI developed in that work is applicable to underdoped SC and inapplicable to describe experiments with overdoped samples which were used in Božović et al. (2016). In particular, in underdoped samples, we cannot expect a linear dependence of the critical temperature on the density of SC electrons which was observed in Božović et al. (2016). This dependence should rather be expected to be nonlinear, as it follows from eq. (4.5.3) of Chapter 4.

To describe the overdoped regime, a theory (Shaginyan et al., 2017) has recently been constructed on the basis of Fermi condensation described in Dukelsky et al. (1997). It is a generalization of the BCS theory where it was shown that the number of SC current carriers is only a small portion of their total number which is in agreement with the results of Božović et al. (2016).

https://doi.org/10.1515/9783110786668-005

Hence, we can conclude that the results obtained in Božović et al. (2016) are rather general and are valid for both underdoped and overdoped cases (see also Božović et al., 2017).

We can also expect that the temperature dependence of the resistance is linear for $T > T_c$ in the underdoped and overdoped cases since the number of bipolarons is small as compared to the total number of electrons, if EPI is dominant and a crystal is isotropic.

In contrast to Shaginyan et al. (2017) in recent work of Pashitskii (2016) it was shown that the linear dependence of T_c on the number of Cooper pairs which was observed in Božović et al. (2016) for overdoped $La_{2-x}Sr_xCu_2O$ crystals can be explained in terms of the BCS on the basis of plasmon mechanism of SC. Nevertheless, it seems that the special case considered in Pashitskii (2016) cannot explain the general character of the results obtained in Božović et al. (2016).

The problem of inability of the BCS and other theories to explain the results of Božović et al. (2016) was also considered in recent work (Hai et al., 2018) where a simple model of a bipolaron SC is developed and the number of bipolaron current carriers is shown to be small as compared to the total number of electrons. The results obtained in Hai et al. (2018) confirm the results of Lakhno (2018, 2019a, 2019b) that the portion of paired states is small in the low-temperature limit.

Important evidence in favor of bipolaron mechanism of SC is provided by experiments on measuring the noise of tunnel current in LSCO/LCO/LSCO heterostructures performed in Zhou et al. (2019).

According to these experiments, paired states of current carriers exist at $T > T_c$ too, that is, they form before the formation of a superconducting phase. This crucially confirms the applicability of the bipolaron scenario to high-temperature oxides. This conclusion is also confirmed by the results of terahertz spectroscopy (Bilbro et al., 2011).

Figure 4.4 illustrates typical dependencies of $E(\tilde{T})$. They suggest that at the transition point the energy is a continuous function $\tilde{T}$. This means that the transition per se proceeds without expending energy and the transition is the second-order phase transition in full agreement with the experiment. At the same time the transition of Bose particles from the condensed state to the supracondensed one proceeds with consuming energy which is determined by quantity q (Section 4.3, Table 4.1) which determines latent transition heat of Bose gas, therefore the first-order phase transition takes place.

Let us consider $YBa_2Cu_3O_7$ (YBCO) HTSC with the transition temperature 90–93 K, the unit cell volume $0.1734 \times 10^{-21} cm^3$, and hole concentration $n \approx 10^{21} cm^{-3}$. According to estimates made in Gor'kov and Kopnin (1988), the Fermi energy is equal to $E_F = 0,37$ eV. The concentration of TI bipolarons in $YBa_2Cu_3O_7$ can be found from eq. (4.5.3):

$$\frac{n_{bp}}{n} C_{bp} = f_{\tilde{\omega}}(\tilde{T}_c),$$

with $\tilde{T}_c = 1,6$. Table 4.1 lists the values of $n_{bp,i}$ for $\tilde{\omega}_i$ parameters presented in it. Table 4.1 suggests that $n_{bp,i} \ll n$. Hence, only a small portion of current carriers are in the bipolaron state. It follows that in full agreement with the results of Section 4.4, the Coulomb interaction of bipolarons will be screened by unpaired electrons, which justifies the approximation of a noninteracting TI bipolaron gas considered.

According to this approach, for an SC to arise paired states should form. The condition of the formation of such states in the vicinity of the Fermi surface, according to Lakhno (2017), has the form: $E_{bp} < 0$. Accordingly, the value of the pseudogap, according to the results of Section 4.4, will be:

$$\Delta_1 = \left| E_{bp} + u_0 \right|. \tag{5.1.1}$$

Naturally, this value is independent of the vector **k**, but depends on the concentration of current carriers, that is, the level of doping.

In the simplest version of the SC theory under consideration, the gap ω_0 does not change in passing on from the condensed to the noncondensed state, that is, in passing on from the superconducting to the nonsuperconducting state and, therefore, ω_0 has also the meaning of a pseudogap:

$$\Delta_2 = \omega_0(\mathbf{k}), \tag{5.1.2}$$

which depends on the wave vector **k**.

Numerous discussions on the gap and pseudogap problem stem from the statement that the energy gap in HTSC is determined by the coupling energy of Cooper pairs which leads to insoluble contradictions (see reviews by Damascelli et al., 2003; Norman et al., 2005; Lee, 2014; Hashimoto et al., 2014; Timusk and Statt, 1999).

Actually, the value of a SC gap Δ_2, determined by (5.1.2), generally speaking, does not have anything to do with the energy of paired states which is determined by E_{bp}. According to Lakhno (2016b), for small values of the EPI constant α, and for large ones, the bipolaron energy $\left|E_{bp}\right| \sim \alpha^2 \omega_0$, that is, $\left|E_{bp}\right|$ does not depend on ω_0.

For example, in the framework of the concept considered, it is clear why the pseudogap Δ_2 has the same anisotropy as the SC gap – this is one and the same gap. It is also clear why the gap and the pseudogap depend on temperature only slightly. In particular, it becomes understandable why in the course of a SC transition a gap arises immediately and does not vanish at $T = T_c$ (this is not BCS behavior). Much-debated question of what order parameter should be put into correspondence to the pseudogap phase (i.e., whether the pseudogap phase is a special state of the matter (Norman et al., 2005)) seems to be senseless within the theory presented.

Presently, there are a lot of methods for measuring a gap: angle resolved photoemission electron spectroscopy (ARPES), Raman (combination) spectroscopy, tunnel scanning spectroscopy, magnet neutron scattering, and so on. According to Timusk and Statt (1999), for the maximum value of the gap in YBCO (6.6) (in the antinodal direction in the ab-plane) it was obtained $\Delta_1 / T_c \approx 16$. This yields $\left|E_{bp}\right| \approx 80$ meV.

Let us determine the characteristic energy of phonons responsible for the formation of TI bipolarons and superconducting properties of oxide ceramics, that is, the value of a SC gap Δ_2. To do so, we compare the calculated jumps of the heat capacities with the experimental values.

A theoretically calculated in Chapter 4 jump of the heat capacity (Fig. 4.5), coincides with the experimental values of jumps in $YBa_2Cu_3O_7$ (Overend et al., 1994) for $\tilde{\omega} = 1.5$, that is, for $\omega = 7.5$ meV. This corresponds to the TI bipolaron concentration $n_{bp} = 2.6 \times 10^{18}\text{cm}^{-3}$. Taking into account that $|E_{bp}| \approx 0.44\alpha^2\omega$ (Lakhno, 2015b), $|E_{bp}| = 80$ meV, $\omega = 7.5$ meV, the EPI constant will be: $\alpha \approx 5$, which is far beyond the limits of the BCS applicability.

As is known, in the BCS theory a jump of the heat capacity is equal to: $(C_S - C_n)/C_n = 1.43$ (where C_S is the heat capacity in the superconducting phase, and C_n is that in the normal one) and does not depend on the model Hamiltonian parameters. As it follows from numerical calculations presented in Fig. 4.5 and Table 4.1 in Chapter 4, as distinct from the BCS, the jump value depends on the phonon frequency. Hence, the approach presented predicts the existence of the isotope effect for the heat capacity jump.

It should be noted that in calculations of the transition temperature it was believed that the effective mass M_e in eq. (4.5.3) is independent of the wave vector direction, that is, an isotropic case was considered.

In an anisotropic case, choosing the main axes of vector **k** for the coordinate axes, we will get $(M_{ex}M_{ey}M_{ez})^{1/3}$ instead of the effective mass M_e. In layered HTSC materials the values of effective masses lying in the plane of the layers M_{ex}, M_{ey} are close in value. Assuming that $M_e = M_{ex} = M_{ey} = M_{||}$, $M_{ez} = M_{\perp}$, we get instead of C_{bp}, determined in (4.5.3), the quantity $\tilde{C}_{bp} = C_{bp}/\gamma$, where $\gamma^2 = M_{\perp}/M_{||}$ is the anisotropy parameter. Hence consideration of the anisotropy of effective masses gives for concentration n_{bp} the value $\tilde{n}_{bp} = \gamma n_{bp}$. Therefore, consideration of anisotropy can enlarge the estimate of the TI bipolaron concentration by an order of magnitude and greater. If for $YBa_2Cu_3O_7$ we take the estimate $\gamma^2 = 30$ (Marouchkine, 2004), then for the TI bipolaron concentration we get: $\tilde{n}_{bp} = 1.4 \cdot 10^{19}\text{cm}^{-3}$, which leaves in place the main conclusion: in the case considered only a small portion of current carriers are in the TI bipolaron state. The situation can change if the anisotropy parameter is very large. For example, in layered HTSC Bi-Sr-Ca-Cu-O the anisotropy parameter is $\gamma > 100$, therefore the concentration of TI bipolarons in these substances can be of the same order as the concentration of current carriers.

Another important conclusion suggested by consideration of anisotropy of effective masses is that the transition temperature T_c depends not on n_{bp} and $M_{||}$ individually, but on their relation which straightforwardly follows from (4.5.3) in Chapter 4. This phenomenon is known as Uemura law. In the Section 4.10, we discussed a more general relation, known as Alexandrov formula (for which Uemura law is a particular case).

Among the experiments involving an external magnetic field, those measuring the London penetration depth λ are of great importance. In $YBa_2Cu_3O_7$ for λ at $T=0$ in Buckel and Kleiner (2004) it was obtained that $\lambda_{ab}=150-300\text{nm}$, $\lambda_c=800\text{nm}$. The same order of magnitude is obtained for these quantities in a number of works (Edstam and Olsson, 1994; Panagopoulos et al., 1998; Pereg-Barnea et al., 2004; Bonn et al., 1993). In Pereg-Barnea et al. (2004) (see also references therein), it is shown that anisotropy of depths λ_a and λ_b in cuprate planes can account for 30% depending on the type of the crystal structure. If we take the values $\lambda_a=150\text{nm}$ and $\lambda_c=800\text{nm}$, obtained in most papers then the anisotropy parameter, according to (4.7.9) in Chapter 4, will be $\gamma^{*2}\approx 30$. This value is usually used for $YBa_2Cu_3O_7$ crystals.

The temperature dependence $\lambda^2(0)/\lambda^2(T)$ was studied in many works (see Bonn et al., 1993 and references therein). Figure 5.1 compares the different curves for $\lambda^2(0)/\lambda^2(T)$. In Bonn et al. (1993), it was shown that in high-quality $YBa_2Cu_3O_7$ crystals the temperature dependence $\lambda^2(0)/\lambda^2(T)$ is well approximated by a simple dependence $1-t^2$, $t=T/T_c$. Figure 5.2 illustrates a comparison of the experimental dependence $\lambda^2(0)/\lambda^2(T)$ with the theoretical one:

$$\frac{\lambda^2(0)}{\lambda^2(T)}=1-\left(\frac{T}{T_c}\right)^{3/2}\frac{F_{3/2}(\omega/T)}{F_{3/2}(\omega/T_c)}, \tag{5.1.3}$$

which results from (4.7.10), (4.5.5), and (4.5.6) in Chapter 4. Hence, there is a good agreement between the experiment and the theory (5.1.3).

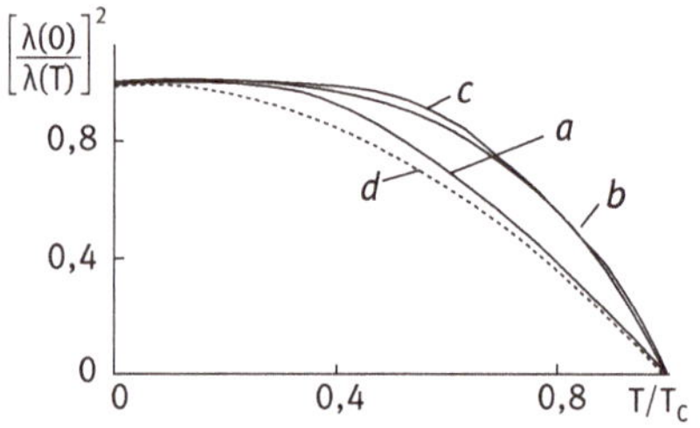

Fig. 5.1: Depth of a magnetic field penetration in the BCS theory (a – local approximation, b – nonlocal approximation); empirical rule $\lambda^{-2}\sim 1-(T/T_c)^4(c)$ (Madelung, 1972); $YBa_2Cu_3O_7$ (d) (Bonn et al., 1993).

The theory developed enables one to compare the temperature dependence of the critical magnetic field in $YBa_2Cu_3O_7$ (Wu and Sridhar, 1990). Since the theory developed in Section 4.8 describes a homogeneous state of a TI bipolaron gas, the critical field being considered corresponds to a homogeneous Meissner phase. In Wu and Sridhar (1990) this field is denoted as H_{c1} which relates to the denotation of Section 4.8 as: $H_{c1}=H_{cr}$, $H_{c1\parallel}=H_{cr\perp}$, $H_{c1\perp}=H_{cr\parallel}$. For comparison with the experiment, we use the parameter values earlier obtained for $YBa_2Cu_3O_7$: $\tilde{\omega}=1.5$, $\tilde{\omega}_c=1.6$. Figure 5.3 compares the experimental dependencies $H_{c1\perp}(T)$ and $H_{c1\parallel}(T)$ (Wu and Stridhar, 1990) with theoretical ones (4.8.13), where for $H_{\max\parallel,\perp}(T)$ experimental values $H_{\max\parallel}(T)=240$, $H_{\max\perp}(T)=816$ are taken. The results shown in Fig. 5.3 confirm the conclusion (Section 4.8) that the relations $H_{cr\perp}(T)/H_{cr\parallel}(T)$ are independent of temperature.

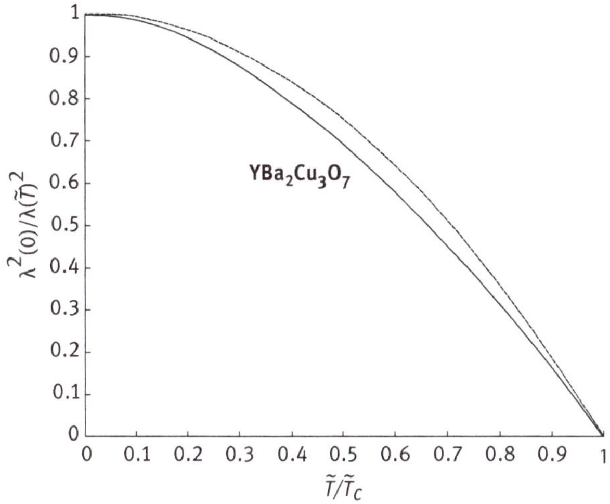

Fig. 5.2: Comparison of the theoretical dependence $\lambda^2(0)/\lambda^2(T)$ (solid curve), obtained in Lakhno (2019b), with the experimental one (Bonn et al., 1993) (dotted curve).

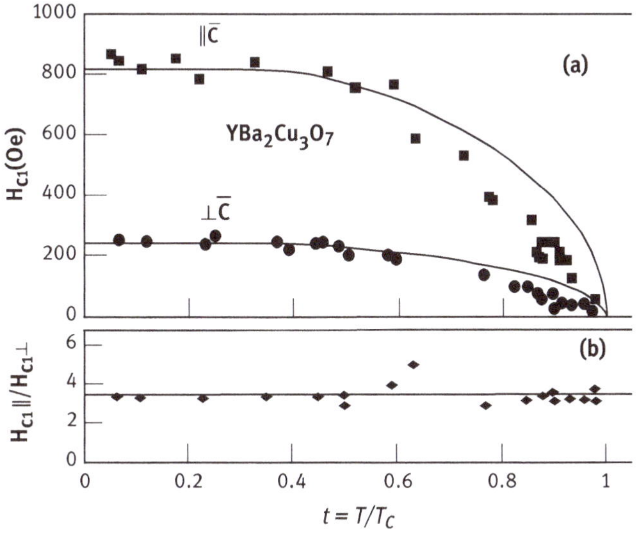

Fig. 5.3: Comparison of calculated (continuous curves) and experimental values of H_{c1}(squares, circles, rhombs; Wu and Sridhar, 1990) for the cases $\|c$ and $\perp c$.

It follows from relations (4.7.9), (4.8.10), and (4.11) in Chapter 4 that:

$$(\gamma^*)^2 = \frac{M_\perp^*}{M_\|^*} \propto \frac{\lambda_\perp^2}{\lambda_\|^2}; \qquad \frac{H_{\max\perp}^2}{H_{\max\|}^2} = \gamma^2 = 11.6. \tag{5.1.4}$$

The choice of the value $\gamma^2 = 11.6$, determined by (5.1.4), for the anisotropy parameter differs from the value $(\gamma^*)^2 = 30$ used above. This difference is probably caused by a difference in the anisotropy of the polaron effective masses $M^*_{\parallel,\perp}$ and electron band masses.

The presence of a gap ω_0 in HTSC ceramics is proved by numerous spectroscopic experiments (ARPES) on angular dependence of ω_0 on $\boldsymbol{\kappa}$ for small $|\mathbf{k}|$ (Damascelli et al., 2003; Norman et al., 2005; Lee, 2014; Hashimoto et al., 2014; Timusk and Statt, 1999). The availability of d-symmetry in the angular dependence $\omega_0(\mathbf{k})$, is probably concerned with the appearance of a pseudogap and transformation of Fermi system into the system of Fermi arcs possessing d-symmetry. In experiments on tunnel spectroscopy the quantity ω_0 can manifest itself as an availability of a gap substructure superimposed on pseudogap $\Delta_1(\Delta_1 \gg \omega_0)$. This structure was frequently observed in optimally doped $YBa_2Cu_3O_7$ and $Bi_2Sr_2CaCu_2O_8$ (BCCO) in the range of 5–10 meV (Maggio-Aprile et al., 1995; Pan et al., 2000; Hoogenboom et al., 2000), which coincides with the estimate of ω_0 presented above.

In a lot of experiments the dependence of the gap and pseudogap value on the level of doping x is measured. Even early experiments on magnetic susceptibility and Knight shift revealed the availability of the pseudogap which emerges for $T^* > T_c$. Numerous subsequent experiments revealed the peculiarities of the $T - x$ phase diagram: T^* increases and T_c decreases as doping decreases (Damascelli et al., 2003; Norman et al., 2005; Lee, 2014; Hashimoto et al., 2014; Timusk and Statt, 1999). As it is shown in Lakhno (2017), this behavior can be explained by peculiarities of the existence of bipolarons in a polaron gas.

It is noted in Lakhno (2017) that 1/8 anomaly (Fig. 5.4) in HTSC systems (Schrieffer, 2007) has probably general character.

The stability condition $E_{bp} < 0$ presented above means that the presence of Fermi gas radically changes the criterion of bipolaron stability which, in the absence of Fermi environment, takes on the form $E_{bp} < 2E_p$. This stabilization was first pointed out in Shanenko et al. (1996) and Smondyrev et al. (2000). This fact plays an important role in explaining concentration dependencies of T_c on x. Most probably, in real HTSC materials the value of the EPI constant has an intermediate value. Then in the range of small concentrations in the absence of Fermi environment, TI bipolarons are unstable with respect to their decay into individual polarons and SC at small x is impossible. It arises for finite x when there is a pronounced Fermi surface which stabilizes the formation of bipolarons (see Chapter 7). This corresponds to a lot of experiments on HTSC materials. A simple thermodynamic analysis (Chapter 3) demonstrates that at a finite temperature TI bipolarons are stable if: $|E_{bp} - 2E_p| \geq T$. Hence, the characteristic temperature T^*, corresponding to the pseudogap phase is equal to: $T^* \approx |E_{bp} - 2E_p|$.

The transition to the pseudogap phase per se is concerned with the formation of TI bipolarons for $T < T^*$ and highly blurred with respect to temperature in full agreement with the experiment. It should be noted that $T^* \ll |E_{bp}|$ where T^* approximately 1,5–2 times exceeds T_c.

As doping increases at $x > x_{opt}$, where x_{opt} is the value of optimal doping, SC passes on to overdoped regime when the number of bipolarons becomes so large that they start overlapping, that is, a transition to the regime of BCS with small T_c takes place.

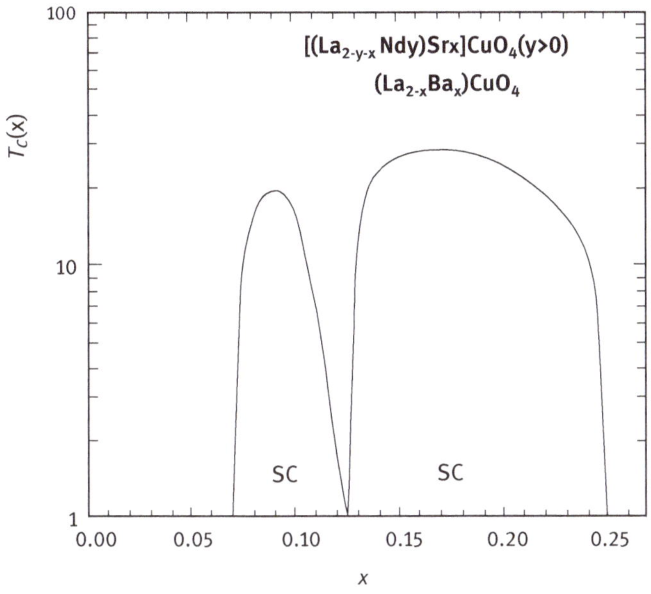

Fig. 5.4: Dependence of T_c (x) for high-temperature superconductors with 1/8 anomaly.

In conclusion it should be noted that the long-term discussion of the nature of the gap and pseudogap in HTSC materials is largely related to the methodological problem of measurements when different measuring techniques actually measure not the same but different quantities. In the case under consideration ARPES measures $\omega_0(\mathbf{k})$, while tunnel spectroscopy – $|E_{bp}|$. Below we consider these problems in greater detail.

5.2 Spectroscopic experiments

As it is shown in the previous section, the theory developed is consistent with thermodynamic and magnet characteristics of HTSC materials. However, these facts are insufficient to judge unambiguously that the TI bipolaron theory of SC does not contradict other experimental facts.

Presently there are a lot of methods to study the properties of paired states and consequences of these states. The aim of this section is to analyze to what extent the data of modern spectroscopic methods such as scanning tunnel microscopy (STM), quasiparticle interference, angle-resolved photoelectron spectroscopy (ARPES), and Raman (combination) scattering are compatible with the ideas of the TI bipolaron mechanism of HTSC.

5.2.1 Tunnel characteristics

In the case of the TI bipolaron theory of SC tunnel characteristics have their peculiarities. As usual, in considering tunnel phenomena, for example, in considering a Josephson transition from a superconductor to an ordinary metal via a tunnel contact we will reckon the energy from the ground state of the SC. In the TI bipolaron theory of an SC, the ground state is the bipolaron state whose energy is below the Fermi level of this SC in the normal state by the value of the bipolaron energy $|E_B|$. Hence, as a result of a tunnel contact of a SC with a conventional metal the Fermi level of a conventional metal will coincide with the ground-state energy of a SC. It follows that the one-particle current will have the usual form for such a contact (Fig. 5.5).

A peculiarity arises in considering a two-particle current. It is concerned with the fact that the spectrum of excited states of a TI bipolaron is separated from the ground state by the value of the phonon frequency ω_0. For this reason, the volt–ampere characteristic of a two-particle current will have the form shown in Fig. 5.5, where $|E_B/2|$ is replaced by ω_0. As a result, the resulting volt–ampere characteristic will have the form of Fig. 5.6.

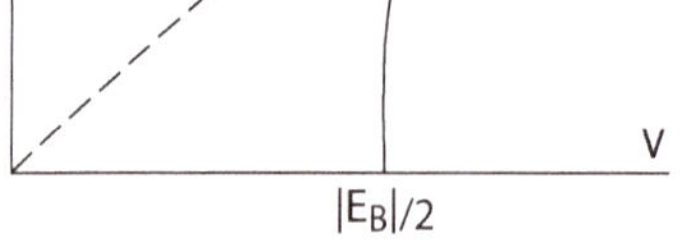

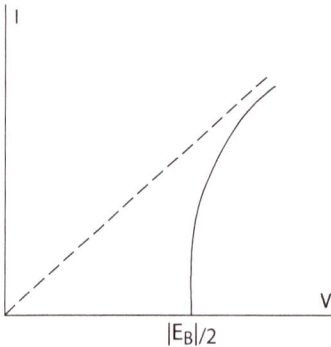

Fig. 5.5: Volt–ampere characteristic of a one-particle current.

The curve I–V is constructed for the case of $\omega_0 < |E_B|/2$. In the opposite case, the quantities ω_0 and $|E_B|/2$ should be inverted. The $\omega_0 < V < |E_B|/2$ segment of the I–V curve in Fig. 5.6, corresponds to a kink which is lacking in the BCS theory.

Spectrally, a kink corresponds to a transition of a one-particle electron spectrum with energy lying lower than E_F by the value of $|E_B|/2$, to a two-particle TI bipolaron spectrum of excited states which in a one-particle scheme lies in the range of $(E_F - |E_B/2| + \omega_0/2, E_F)$, as it is shown in Fig. 5.7.

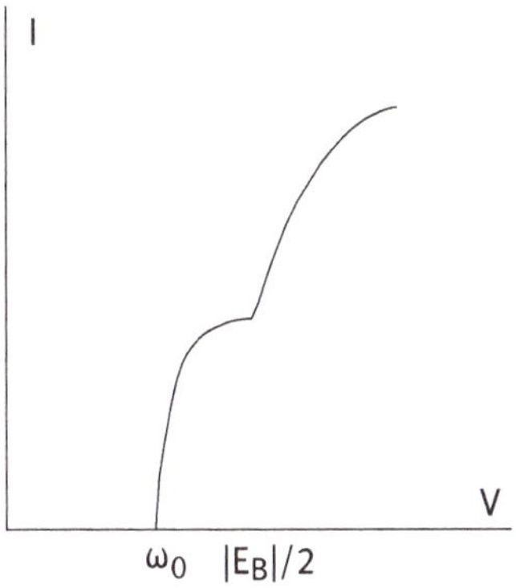

Fig. 5.6: Volt–ampere characteristic of a total current.

The dependence $E(k)$ shown in Fig. 5.7 corresponds to ARPES observations of kinks in a lot of HTSC materials (e.g., review of Garcia and Lanzara, 2010). For example, according to Garcia and Lanzara (2010) in the well-studied cuprate Bi2212 the kink energy ($|E_B|/2$) is 70 meV.

The phonon nature of the kink is also supported by the observation of the isotope effect near the kink energy (Iwasawa et al., 2008), the independence of the kink energy from the doping value (Zhou et al., 2003), and the independence of the kink energy from the nature of current carriers: according to Park et al. (2008), electron- and hole-doped cuprates have the same kink energy.

Figure 5.8. shows a dependence of dI/dV on V, typical for HTSC which corresponds to the dependence of I on V presented in Fig. 5.7. There a kink corresponds to a dip on the curve to the right of the high peak.

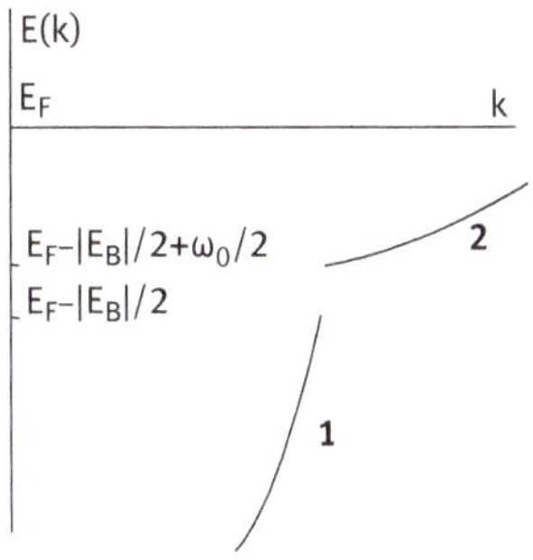

Fig. 5.7: Kink corresponds to a gap in passing on from normal branch 1 to TI bipolaron branch 2 for energy $E_F - |E_B/2|$.

Notice that since TI bipolarons exist for $T > T_c$ too, at temperature exceeding the critical one the dI/dV curve will qualitatively retain the form shown in Fig. 5.8. Hence, the quantity $|E_B|/2$ will play the role of a pseudogap in one-particle transitions, while $|E_B|$ plays the role of a pseudogap in two-particle transitions. This conclusion is in full agreement with numerous tunnel experiments in HTSC (Garcia and Lanzara, 2010; Giubileo et al., 2002; Giubileo et al., 2001).

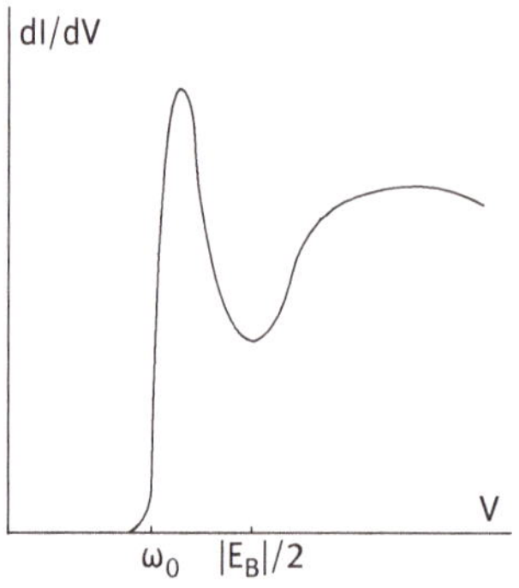

Fig. 5.8: Dependence of conductance dI/dV on V, corresponding to the volt–ampere characteristic shown in Fig. 5.7.

5.2.2 Angle-resolved photoelectron spectroscopy (ARPES)

Apart from STM, a direct method providing information on the properties of a superconducting gap is angle resolved photo electron spectroscopy (Damascelli et al., 2003). Being added by the data of STM and the results of the quasiparticle interference, this method provides the most complete data on the properties of a SC gap. Recently, a method of double photoelectron spectroscopy has been developed where two electrons with certain momenta $\mathbf{k}_1$ and $\mathbf{k}_2$ and relevant energies E_1 and E_2 emit (Hattass et al., 2008). Despite the abundance of data obtained by ARPES the nature of a HTSC gap is still unclear. To a large extent this is due to the fact that up to the present time a unified theory of HTSC was lacking. If we proceed from the fact that a SC mechanism is caused by Cooper pairing, then in the case of strong EPI, this leads to the TI bipolaron theory if HTSC is being considered. According to this theory, as distinct from bipolarons with broken symmetry, TI bipolarons are spatially delocalized and the polarization potential well is lacking (polarization charge is zero). According to Section 4.3 of Chapter 4, a TI bipolaron has a gap in the spectrum which has a phonon nature. In the TI bipolaron theory of SC, bipolarons are formed in the vicinity of the Fermi surface in the form of a charged Bose gas (immersed into electron gas) which condenses at the level lying lower the Fermi level by the value equal to the bipolaron ground-state energy which leads to SC state. The spectrum of excitations of such a gas has a gap equal to the phonon frequency. In this section, we will show that the photoemission spectrum obtained in ARPES just contains this gap and the gap $|E_B|/2$, determined from the two-particle current by STM which was considered in the previous section has nothing to do with the measurements of a gap by ARPES.

To this end, we will proceed from the general expression for the light absorption intensity $I(\mathbf{k}, \omega)$ measured in ARPES in the form:

$$I(\mathbf{k}, \omega) = A(\mathbf{k}, \omega)F(\omega)M(\mathbf{k}, \omega). \tag{5.2.1}$$

In the case of the intensity of light absorption by TI bipolarons measured by ARPES, the quantities involved in (5.2.1) have a different meaning than in the case of a one-electron emission.

In the case of a Bose condensate considered k has the meaning of a Boson momentum and ω is boson energy, $A(\mathbf{k},\omega)$ is a one-boson spectral function, $F(\omega)$ is a Bose–Einstein distribution function, $M(\mathbf{k},\omega)$ is a matrix element which describes transitions from the initial boson state to the final one.

In our case, the role of a charged boson taking part in the light absorption belongs to a bipolaron whose energy spectrum is determined by (4.3.1) and (4.3.2) of Chapter 4:

$$\varepsilon_k = E_B \Delta_{k,0} + \left(E_B + \omega_0(\mathbf{k}) + k^2/2M\right)\,(1 - \Delta_{k,0}), \tag{5.2.2}$$

where $\Delta_{k,0} = 1$, if $k = 0$, $\Delta_{k,0} = 0$, if $k \neq 0$, whose distribution function is $F(\omega) = [\exp(\omega - \mu) - 1]^{-1}$. For $\mathbf{k} = 0$ TI bipolaron is in the ground state, while for $\mathbf{k} \neq 0$ – in the excited state with energy $E_B + \omega_0(\mathbf{k}) + k^2/2M$, where $\omega_0(\mathbf{k})$ is a phonon frequency depending on the wave vector, $M = 2m$, m is the electron effective mass.

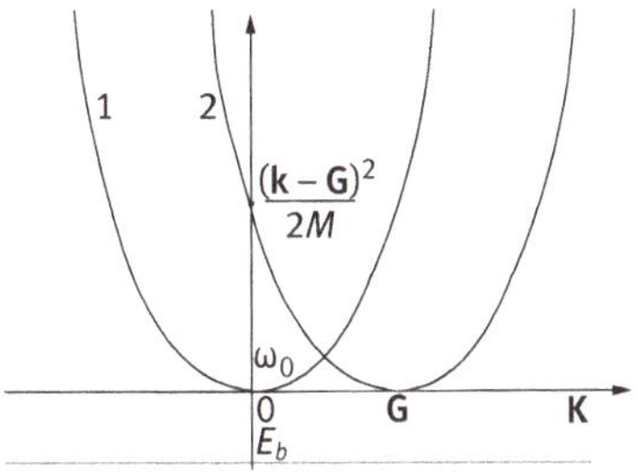

Fig. 5.9: Schematic representation of the bipolaron transition to the excited state as a result of absorption of light quanta.

For further analysis, it should be noted that the energy of bipolaron excited states reckoned from E_B in eq. (5.2.2), can be interpreted as the energy of a phonon $\omega_0(\mathbf{k})$ and the kinetic energy of two electrons coupled with this phonon. The latter, in the scheme of extended bands, has the form: $(\mathbf{k} + \mathbf{G})^2/2M$, where $\mathbf{G}$ is the lattice inverse vector (Fig. 5.9). ARPES measures the spectrum of initial states which in our case is the spectrum of low-lying excitations of a TI bipolaron. In this connection we can neglect the contribution of one- and two-particle excitations of the electron (polaron) gas into which the bipolarons are immersed since the density of the TI bipolaron states in the vicinity of their ground state is much greater than that of the electron spectrum states. Hence, we a priori exclude consideration of such phenomena as the de Haas–van Alphen oscillation and the Shubnikov–de Haas oscillation (Vignolle et al., 2008; Yelland et al., 2008; Helm et al., 2009). Since the kinetic energy corresponding to the inverse lattice vector (or the whole number of the inverse lattice vectors) is very high, out of whole spectrum of a bipolaron determined by (5.2.2), we should take account only of the levels E_B with $k = 0$ and $E_B + \omega_0(\mathbf{k})$ with $k \neq 0$ as a spectrum of the initial states. In other words, with the use of the spectral function $A(\omega, \mathbf{k}) = -(1/\pi)\mathrm{Im}\, G(\omega, \mathbf{k})$, where $G(\omega, \mathbf{k}) = (\omega - \varepsilon_k - i\varepsilon)^{-1}$ is the Green bipolaron function, the expression for the intensity (5.2.1) can be presented as follows:

$$I(\mathbf{k}, \omega) \sim \left((\omega - E_B)^2 + \varepsilon_1^2\right)^{-1} \cdot \left((\omega - E_B - \omega_0(\mathbf{k}))^2 + \varepsilon_2^2\right)^{-1}, \tag{5.2.3}$$

which is fitting of the distribution function F with $\mu = E_B$ and Green function G by Lorentzians where ε_1 and ε_2 determine the width of the Bose distribution and bipolaron levels, respectively (matrix element $M(\mathbf{k}, \omega)$, involved in (5.2.1), has a smooth dependence on the energy and wave vector, therefore this dependence can be neglected).

Hence, as a result of light absorption by a pair of electrons (which are initially in a bipolaron state), ARPES measures the kinetic energy of electrons with momenta k_e, which are expelled from the sample in vacuum as a result of absorption of a photon with energy $\hbar v$. The energy conservation law in this case takes on the form:

$$\hbar v = \omega_0(\mathbf{k}) + \frac{(\mathbf{k} + \mathbf{G})^2}{2M} = \xi + \frac{k_e^2}{m_0},$$
$$\xi = 2\Phi_0 + |E_B|, \tag{5.2.4}$$

which is illustrated by Fig. 5.10, where Φ_0 is the work of electrons escape from the sample, m_0 is the mass of a free electron in vacuum. Figure 5.10 suggests that when a bipolaron is formed in the vicinity of the Fermi energy E_F the energy of two electrons becomes equal to $2E_F + E_B$.

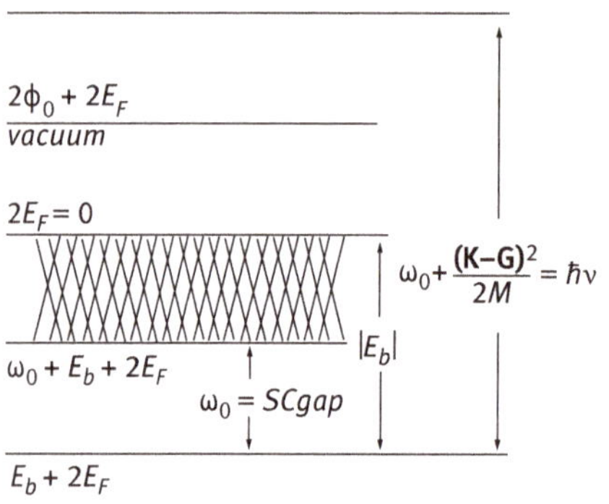

Fig. 5.10: Scheme of energy levels in measuring the spectrum by ARPES. The region of the continuous spectrum lying below the Fermi level is shaded.

In this case, electrons pass on from the state with p_F, where p_F is the Fermi momentum, to a certain state with momentum p below the Fermi surface (since $E_B < 0$). ARPES measures the spectrum of initial states reckoned from the energy $2 \cdot p^2/2m$, which corresponds to the energy of two electrons with momentum p. As a result, ARPES measures the energy $\omega_i = 2E_F + E_B - p^2/m$.

Hence, if a bipolaron with energy $\omega = \omega_i = E_B + 2v_F(p_F - p)$, lying in domain of existence of a bipolaron gas $(2E_F + E_B, 2E_F)$, where v_F is the velocity of a Fermi electron, absorbs a photon with energy $\hbar\nu$, then a phonon arising as a result of the bipolaron decay recorded in ARPES as a gap $\omega_0(\mathbf{k})$, and two electrons with the kinetic energy k_e^2/m_0, determined by (5.2.4) are emitted from the sample.

In this scenario in each act of light absorption two electrons with similar momenta are emitted from the sample. This phenomenon can be detected by ARPES if the electron detector is placed just on the sample surface since the kinetic energy of flying of the emitted electron pair in vacuum (not compensated by the attracting potential in the bipolaron state) is several electronvolts.

Hence, ARPES, as discussed above, measures the phonon frequency $\omega_0(k)$, which is put into correspondence to the SC gap and therefore in cuprate HTSC with $d_{x^2-y^2}$ symmetry its angular dependence is determined by the expression $\omega_0(\mathbf{k}) = \Delta_0 \left|\cos k_x a - \cos k_y a\right|$.

From the viewpoint of phonon spectroscopy, identification of phonon modes of this type is difficult in view of their small number (equal to the number of bipolarons) as compared to the number of ordinary phonons equal to the number of atoms in a crystal. The spectral dependence of phonon frequencies is determined by both ion–ion interactions and an interaction with the electron subsystem of the crystal. Calculation of normal oscillations for a plane square lattice of atoms without taking account of the electron contribution leads to *d*-symmetry of their spectrum (Emin, 2017; Okomel'kov, 2002). With regard to CuO_2 SC planes of oxide ceramics, in the direction of Cu–O–Cu bonds (antinodal direction), phonons will have a gap, while in the direction of Cu–Cu bonds, that is, along the unit cell diagonal (nodal direction) a gap will be lacking.

In calculating the electron contribution into the phonon spectrum account should also be taken of the relation between the electron density distribution and the position of ions on CuO_2 plane observed in STM/STS experiments with high spatial resolution (Lawler et al., 2010).

The angular dependence $\omega_0(k)$ leads to the angular dependence of the intensity $I(\omega_i, \mathbf{p}) \sim A(\omega_i, \mathbf{p})$, determined by eq. (5.2.3) (Fig. 5.11), which is usually observed in ARPES experiments (Damascelli et al., 2003; Borisenko et al., 2001; Shen et al., 2004). The form of the $(\omega_i, \mathbf{p})$ dependence suggests that there is also a dependence of the absorption peaks on $\mathbf{p}$ symmetric about the Fermi level. This dependence is not presented in Fig. 5.11, since in view of a small population density of states with $p > p_F$ their absorption intensity will be very small (Matsui et al., 2003).

Experimental checking of the effect of TI bipolaron emission as a whole is important for understanding the pairing mechanisms. Thus, according to Shen et al. (2004), only one electron should escape from the sample with dispersion of the initial states determined, for $P \neq 0$ by the formula: $\varepsilon_p^{Bog} = \sqrt{(p^2/2m - E_F)^2 + \Delta^2(p)}$ where ε_p^{Bog} is the spectrum of a Bogolyubov quasiparticle, different from spectrum (5.2.2).

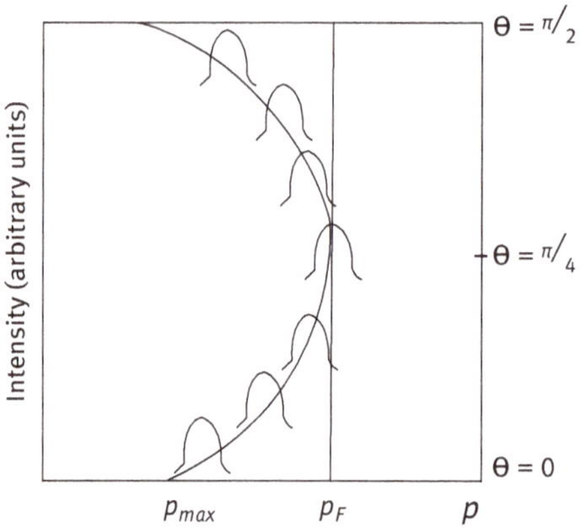

Fig. 5.11: Schematic representation of the angular dependence of the absorption intensity determined by (5.2.3) for $\omega = \omega_i$.

The use of spectra ε_p^{Bog} and (5.2.2) to describe the angular dependence of the intensity leads to a qualitative agreement with the ARPES data with currently accessible resolution. Experiments with higher resolution should give an answer to the question of whether a SC condensate in cuprates has fermion or TI bipolaron character.

The spectrum $\omega_0(k)$ suggests that in cuprate superconductors EPI constant becomes infinite in the nodal direction. Hence, for bipolarons, a regime of strong coupling takes place in this case. Figure 5.12 shows a typical dependence of the absorption intensity $I(\omega_i, \mathbf{p})$, observed in ARPES experiments (Borisenko et al., 2001).

The dependence shown in Fig. 5.12 is obtained from the expression for the intensity (5.2.2) where the spectral function corresponds to the TI bipolaron spectrum (5.2.3) which cannot be obtained from spectral function (5.2.3) from Matsui et al. (2003), where Bogolyubov spectrum ε_p^{Bog} is used for the spectrum and Fermi distribution function is used instead of Bose distribution $F(\omega)$. This result can be considered as an argument in favor of a TI bipolaron mechanism of SC.

The peculiarities of the ARPES absorption spectrum considered above will also manifest themselves in tunnel experiments in the form of a thin structure (kinks) on the volt–ampere characteristics measured. To observe these peculiarities, as distinct from traditional ARPES measurements with high-energy photon sources ($\hbar\nu = 20 - 100$ eV), one should use low-energy photon sources ($\hbar\nu = 6 - 7$ eV) with higher momentum resolution (Vishik et al., 2010; Plumb et al., 2010; Anzai et al., 2010; Rameau et al., 2009).

In Kouzakov and Berakdar (2003), a theoretical possibility to observe the emission of Cooper pairs by ARPES was considered for conventional SC. In particular, Kouzakov and Berakdar (2003) demonstrated the availability of a peak in the emission current of Cooper pairs which corresponds to zero coupling energy of occupied two-electron states. The peak considered in Kouzakov and Berakdar (2003) corresponds

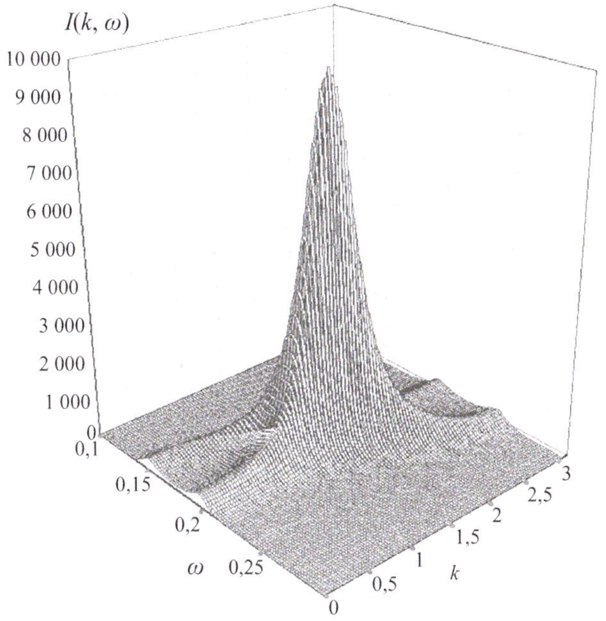

Fig. 5.12: Dependence of the absorption intensity $I(k, \omega)$ (arbitrary units) on k and ω (eV), determined by (5.2.3), for the parameters: $|E_B| = 0.2$eV, $\Delta_0 = 0.05$ eV, $\varepsilon_1 = \varepsilon_2 = 0.01$ eV and the wave vector k in antinodal direction. The lattice constant is assumed to be equal to 1.

to a transition with energy $\hbar\nu$, determined by (5.2.4) where the coupling energy is ~1 meV, which is at the edge of ARPES accuracy. In the case of high-temperature superconductors the coupling energy can be 10 times higher which makes checking of the effects considered more realistic. The main distinction of the results obtained here from those derived in Kouzakov and Berakdar (2003) is the presence of the angular dependence of the absorption peak in Figs. 5.11 and 5.12, which is characteristic for HTSC materials.

Let us briefly discuss the temperature dependence of the intensity $I(\omega_i, \mathbf{p})$. According to eq. (5.2.1), it is determined by the temperature dependence$F(\omega)$.

For $T < T_c$, where T_c is the temperature of a SC transition $F(\omega) \cong N_0(T)$ for $\omega = E_B$, where $N_0(T)$ is the number of bosons (bipolarons) in a condensate which determines the temperature dependence of the absorption intensity. The value of $N_0(T)$ decreases as T grows and, generally speaking, vanishes at the SC transition temperature making the absorption intensity vanish. Actually, however, this is not the case since only the Bose-condensate part vanishes. According to the TI bipolaron theory of SC, for $T > T_c$, bipolarons exist in the absence of a condensate too. In this case, the population density of the ground state of such bipolarons will decrease as the temperature grows vanishing at T^*, which corresponds to a transition from the pseudogap state to the normal one.

This conclusion is confirmed by ARPES experiments in SC and pseudogap phases (Varelogiannis, 1998), which demonstrated that the angular dependence of a d-type SC gap is similar to the angular dependence of the state density in the pseudogap phase. At the same time, there are considerable differences between the ARPES experimental data obtained for the SC phase and the gap one. In the SC phase, the peak of absorption intensity occurs below the Fermi level which corresponds to a sharp spectral peak of the density of Bose-condensate states determined by eq. (5.2.3), while in the pseudogap phase this peak is lacking in view of the lack of a condensate in it (Norman et al., 2005). Under these conditions, since the population density of bipolaron excited states grows with growing temperature, the intensity of the absorption peak in ARPES experiments will decrease with growing temperature and reach minimum in the antinodal direction and maximum in the nodal one.

5.2.3 Neutron scattering

Neutron scattering is a powerful method to study the properties and structure of crystals, widely used in modern physics. With the help of neutrons, one can determine the atomic structure, establish the magnetic structure, obtain information about the spectral characteristics of various excitations in crystals, liquids, and so on.

The neutron scattering method plays one of the key roles in the study of HTSC. In the model of superconductivity, we are considering, based on Bose condensation of TI bipolarons, a neutron incident on the system under study (in our case, such a system is a TI bipolaron condensate) removes it from the state of thermodynamic equilibrium. In this case, the state of the incident neutron itself changes. By registering the energy and momentum of the incident scattered neutron, we obtain information about the excitations that are possible in the system under study.

As is known, the cross section of inelastic scattering σ on a system of particles of the same type at temperature $T = \beta^{-1}$ per unit solid angle Ω has the form:

$$\frac{d^2\sigma}{d\Omega d\omega} = \frac{M_n^2}{(2\pi)^3\hbar^4}\frac{k_f}{k_i}|V(\mathbf{k})|^2 N \frac{S(\mathbf{k},\omega)}{1-e^{-\beta\omega}} \tag{5.2.5}$$

where N is the number of particles in the system, M_n is the neutron mass, $\hbar\mathbf{k} = \hbar(\boldsymbol{k}_i - \boldsymbol{k}_f)$. is the scattering momentum, where $\mathbf{k}_i$ and $\mathbf{k}_f$ are the initial and final neutron momenta, respectively; ω is the dissipated energy; $V(\mathbf{k})$ is the Fourier component of the potential of interaction of a neutron with a scattering particle; $S(\mathbf{k},\omega)$ is a dynamical structure factor related to the correlation function of the density–density type as:

$$S(\mathbf{k},\omega)=\int d^3r\int_{\infty}^{\infty}\frac{dt}{\hbar}e^{-ikr+i\omega t/\hbar}\langle[\rho(r,t),\rho(0,0)]\rangle, \tag{5.2.6}$$

where $\rho(r,t)=\Psi^+(r,t)\Psi(r,t)$ is particle density operator in the Heisenberg representation, $\langle\cdots\rangle$ is the thermodynamic mean.

In the case of an ordinary ideal Bose gas (IBG) whose particles have the spectrum $\varepsilon_q=q^2/2M$ the structure factor $S(\mathbf{k},\omega)$ was calculated in Jackson (1973), Aleksandrov et al. (1975), and Hohenberg and Platzman (1966). In the considered case of an ideal TI bipolaron gas with the spectrum $\varepsilon_0=0$ and $\varepsilon_{k\neq 0}=\omega_0+k^2/2M$, which corresponds to the distribution function $F(\mathbf{k})$:

$$F(\mathbf{k})=n_0(\mathbf{k})\delta(\mathbf{k})+[\exp\beta\varepsilon_k-1]^{-1},\ T<T_c$$
$$F(\mathbf{k})=[\exp\beta(\varepsilon_k-\mu)-1]^{-1},\ T\geq T_c \tag{5.2.7}$$

where $n_0(\mathbf{k})$ is the concentration of TI bipolarons in a Bose condensate, μ is a chemical potential, the expression for the structure factor (5.2.6) takes the form:

$$S(\mathbf{k},\omega)=S_0(\mathbf{k},\omega)+S_1(\mathbf{k},\omega),$$
$$S_0(\mathbf{k},\omega)=n_0\left[(1+F(\mathbf{k}))\delta\left(\omega-\omega_0-k^2/2M\right)-F(\mathbf{k})\delta\left(\omega+\omega_0+k^2/2M\right)\right],$$
$$S_1(\mathbf{k},\omega)=\frac{2\pi M}{\hbar^4\beta k(1-e^{-\beta\omega})}\cdot\ln\frac{1-\exp\left\{-\beta\left[\omega_0+M(\omega+k^2/2M)^2/2k^2-\tilde{\mu}\right]\right\}}{1-\exp\left\{-\beta\left[\omega_0+M(\omega-k^2/2M)^2/2k^2-\tilde{\mu}\right]\right\}},$$
$$\tilde{\mu}=\mu-E_{\text{bp}};\quad \tilde{\mu}=0,\quad T\leq T_c;\quad \tilde{\mu}=\tilde{\mu}(T),\quad T>T_c. \tag{5.2.8}$$

It follows that as $T\to 0$, when $F(\mathbf{k})\to 0$, the processes with energy absorption are lacking and only those with energy transfer are present.

In expression (5.2.8), formfactor $S_0(\mathbf{k},\omega)$determines excitations of individual TI bipolarons in a Bose condensate, while $S_1(\mathbf{k},\omega)$ – a contribution of TI bipolarons occurring in the supracondensate state.

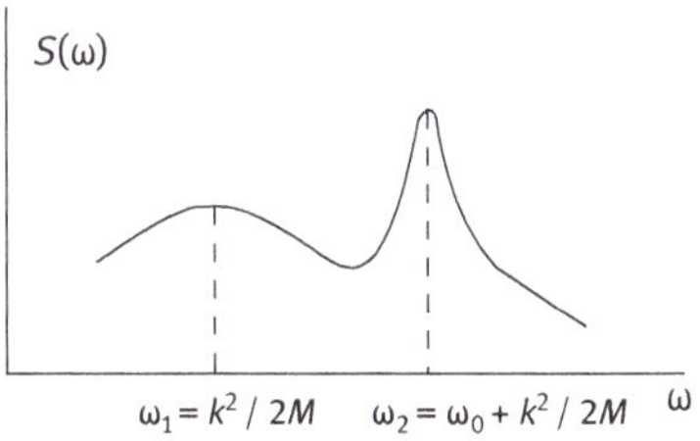

Fig. 5.13: Dependence of a scattering formfactor on the neutron energy.

It follows from (5.2.8) that for $\omega>0$ the formfactor has two peaks. Figure 5.13 shows the function $S(\mathbf{k},\omega)$ for a certain fixed value of $\mathbf{k}$, which has one (blurred) peak for the energy $\omega=\omega_1=k^2/2M$ (it is determined by $S_1(\mathbf{k},\omega)$), and the other (the sharper one) for the energy $\omega = \omega_2 = \omega_0 + k^2/2M$ (it is determined by $S_0(\mathbf{k}, \omega)$) (Fig. 5.13). The energy difference between the two peaks of ω_0, according to the TI bipolaron theory, determined the value of the superconducting gap. For $\omega_0 = 0$ two peaks merge into one which corresponds to IBG.

It follows from (5.2.8) that at $T = T_c$ the maximum for $\omega = \omega_2$ disappears since in this case $n_0 = 0$. The maximum for $\omega = \omega_1$ holds at $T > T_c$ too and disappears only at $T = T^*$, where T^* is the temperature of a transition to the pseudophase. At $T = 0$, on the contrary, the maximum for $\omega = \omega_1$ disappears, while the maximum for $\omega = \omega_2$ becomes maximally sharp and intensive.

The above scenario is in agreement with currently available experiments (Song and Dai, 2015), however, the experimental data are insufficient for an unambiguous conclusion about a phonon or magnon nature of a neutron peak. In a real experiment, a peak in neutron scattering which appears at $T < T_c$ is obtained by subtracting the scattering in the normal and superconducting phases. Hence, the value of an SC gap is put into correspondence to the energy $\omega_2 = \omega_0 + k^2/2M$, since the peak with $\omega = \omega_1$ is automatically subtracted in this approach. This can lead to overestimation of the gap size and a distortion of its angular dependence.

It should be noted that the method of neutron scattering was intensively used to determine the presence of a superfluid component in ^4He, which presumably represents a Bose-condensate state, into which a part of helium atoms transforms at $T < T_c$ (Jackson, 1973; Alexandrov et al., 1975; Hohenberg and Platzman, 1966; Cowley and Woods, 1968; Harling, 1970), as it was first presumed in London (1938).

The theory of neutron scattering by a Bose condensate which is formed at $T < T_c$ and is considered as an ideal gas, leads to the conclusion that there is a sharp peak due to the presence of a condensate, the position of which coincides with the position of the maximum of a wide peak associated with above-condensate excitations (which corresponds to $\omega_0 = 0$). This fact creates great difficulties in separating the Bose-condensate contribution. According to what was said above, this difficulty is absent in the case of a Bose condensate of TI bipolarons, in which the corresponding peaks are separated by the value of ω_0, which makes the inelastic scattering method a promising tool to study the properties of Bose condensates of HTSC materials.

5.2.4 Combination scattering

Though the combination scattering does not provide an angular resolution (Devereaux and Hackl, 2007), its results also testify to the phonon nature of a gap in HTSC. As it was shown in Lakhno (2018, 2019b), the spectrum determined by (5.2.2), can be interpreted as a spectrum of renormalized phonons. Scattering of light with frequency ν on such phonons will lead to an appearance of satellite frequencies $\nu_+^B = \nu + |\varepsilon_k^B|$ and $\nu_-^B = \nu - |\varepsilon_k^B|$ in the scattered light, where ε_k^B is determined by (5.2.2). In the case of wide conductivity bands, that is, when the inequality $G^2/M \gg \max \omega_0(k)$ is fulfilled, split lines $\nu_{\pm}^{B}$ overlap and form a region with a maximum displaced toward the Stocks branch ν_{-}. Since in the model considered the bipolaron gas is placed into the polaron gas where the number of bipolarons is far less than the number of polarons, the intensity of bipolaron satellites will be much weaker than the intensity of TI polaron satellites: $\nu_{+}^{P} = \nu + |\varepsilon_k^P|$ and $\nu_{-}^{P} = \nu - |\varepsilon_k^P|$, $\varepsilon_k^P = E_P\Delta_{k,0} + (\omega_0 + k^2/2m)\cdot(1-\Delta_{k,0})$, E_P is the energy of a TI polaron. As in the case of usual combination scattering, the intensity of scattering on the polarons and bipolarons will be much weaker than the intensity of Rayleigh scattering corresponding to frequency ν.

Indeed, in the combination scattering experiments (Misochko, 2003) at $T < T_C$ a wide peak appears which, according to our interpretation, corresponds to widened frequencies $\nu_{\pm}^{B,P}$. In full agreement with the experiment, the position of this peak is independent of temperature. In the theory of the combination scattering based on the BCS, on the contrary, the position of the peak should correspond to the width of the SC gap and for $T = T_C$ the frequency corresponding to this width should vanish.

The combination scattering results also confirm that TI bipolarons do not decay at $T = T_C$, but persist in the pseudogap phase. Measurement of the temperature dependence of the combination scattering intensity is based on the subtraction of the absorption intensity in the normal and superconducting phases. The difference obtained, according to our approach, is fully determined by scattering on the Bose condensate and depends on temperature vanishing at $T = T_C$.

It should be noted that a lot of spectroscopic experiments are based on the subtraction method. According to TI theory, the result of subtraction of any spectroscopic experiment which measures SC gap in SC phase or in pseudogap phase from the same experiment in normal phase will show the existence of gap. This result can be used to check the TI bipolaron theory of superconductivity.

5.3 Isotope effect

The isotope effect plays a central role in superconductivity. The presence of the isotope effect has played a decisive role in establishing the phonon mechanism of SC in ordinary superconductors. The absence of this effect in optimally doped high-temperature superconductors served as the basis for the rejection of the phonon mechanism in HTSC and, as a consequence, of the BCS theory (Bardeen et al., 1957). In recent years, however, a large number of new experimental facts force us to return to the EPI as the dominant one in explaining the HTSC effect. At the same time, direct use of the BCS theory and its various modifications cannot explain these experimental facts (Bill et al., 1998a, 1998b).

The reason, is probably that the BCS theory, based on EPI, considers this interaction as weak, while in the case of HTSC this interaction turns out to be strong. The generalization of the BCS theory to the case of a strong EPI – the Eliashberg theory, failed to explain many important phenomena accompanying HTSC, for example, the pseudogap state. To overcome these difficulties, in the works of Lakhno (2018, 2019a, 2019b), a TI bipolaron theory of HTSC was constructed, in which the role of Cooper pairs is played by TI bipolarons.

The aim of this section is to explain the isotope effects observed in HTSC on the basis of TI bipolaron theory.

The isotope effect played a decisive role in the establishment of the EPI mechanism of the superconducting state and the substantiation of the BCS theory (Bardeen et al., 1957) for ordinary metals. In the BCS theory, coefficient α is determined from the ratio experimentally established for ordinary metals:

$$T_c M^{\alpha} = const \tag{5.3.1}$$

where T_c is the temperature of a SC transition, M is the mass of an atom replaced by the isotope. It follows from (5.3.1) that:

$$\alpha = -d \ln T_c / d \ln M \tag{5.3.2}$$

In the BCS theory, the value of the coefficient α is positive and close to the value $\alpha \cong 0.5$, which is in good agreement with experiment in ordinary metals. The large value of the isotopic coefficient observed in ordinary metals indicates the dominant role of EPI in them and the applicability of the BCS theory for their description.

On the contrary, in high-temperature superconducting ceramics in the region of their optimal doping, the isotopic coefficient is usually very small $\left(\sim 10^{-2}\right)$, as a result of which EPI in them are small, which necessitated the consideration of other SP mechanisms (Bill et al., 1998a, 1998b).

As is known, the BCS theory developed for the case of a weak EPI is inapplicable in the case of HTSC, in which EPI cannot be considered to be weak. In this case, the TI bipolaron theory of SC can be used (the reasons why the Eliashberg (1960) theory, used in the case of a strong EPI, can be inapplicable for the description of HTSC are discussed in Introduction).

According to TI bipolaron theory, the temperature of SC transition T_c is determined by the equation:

$$\begin{gathered} T_c(\omega_0/T_c) = \left(F_{3/2}(0)/F_{3/2}(\omega_0/T_c)\right)^{2/3} T_c(0) \\ T_c(0) = 3,31\hbar^2 n_{bp}^{2/3}/M_e, \qquad M_e = 2m, \\ F_{3/2}(x) = \frac{2}{\sqrt{\pi}} \int\limits_0^{\infty} \frac{t^{1/2} dt}{e^{t+x} - 1}, \end{gathered} \tag{5.3.3}$$

where n_{bp} is the concentration of TI bipolarons, ω_0 is the frequency of an optical phonon, m is the mass of a band electron (hole), $\hbar = h/2\pi$, h is the Planck constant.

Formulae (5.3.2) and (5.3.3) yield the expression for the isotopic coefficient:

$$\alpha = \frac{0,5}{1+\Phi(y)}$$

$$\Phi(y) = 3y \int_0^\infty \frac{\sqrt{t}dt}{e^{((ty+1)/y)}-1} \Bigg/ \int_0^\infty \frac{dt}{\sqrt{t}(e^{((ty+1)/y)}-1)} \qquad (5.3.4)$$

where $y = T_c/\omega_0$.

The graph of the function $\alpha(T_c/\omega_0)$ is shown in Fig. 5.14. Figure 5.14. suggests that in the case $(T_c/\omega_0) \gg 1$, which can correspond to optimally doped HTSC: high T_c and strong EPI (low ω_0 and large value of EPI constant: $\alpha_{eph}{\sim}\omega_0^{-1/2}$) the isotopic coefficient will be small ($\alpha \to 0$ as $T_c \to \infty$), in full agreement with experiment (Bill et al., 1998a, 1998b; Chen et al., 2007; Franck, 1994; Franck et al., 1991; Batlogg et al., 1987; Zech et al., 1994).

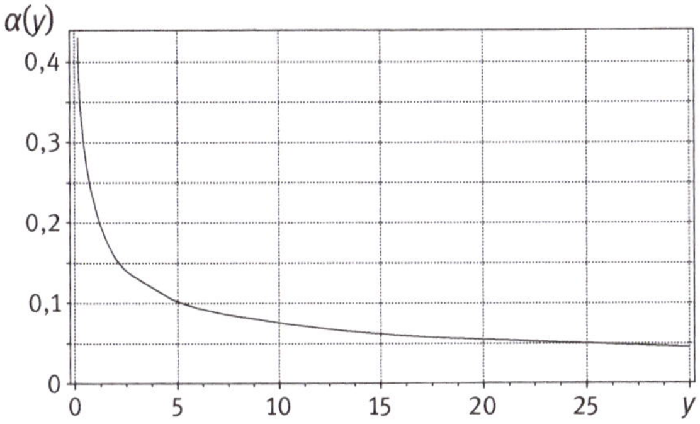

Fig. 5.14: Dependence of isotopic coefficient α on T_c/ω_0.

In the opposite case, $T_c/\omega_0 \ll 1$, which corresponds to the case of weak EPI, the isotopic coefficient reaches its maximum value $\alpha = 0.5$ as in the BCS theory, which corresponds to the case of weak EPI.

It should be noted that according to (5.3.4) the isotopic coefficient of different samples will be the same for the same relation T_c/ω_0.

Figure 5.14 suggests that for typical values of α, the value of phonon frequency ω_0 does not exceed T_c. For HTSC with T_c= 100 K this leads to the value of ω_0 less than 8.6 meV.

Quite a different picture arises for the isotopic coefficient of the London penetration depth:

$$\beta = -\frac{M}{\lambda}\frac{d\lambda}{dM}, \qquad (5.3.5)$$

where λ is London penetration depth:

$$\lambda = \left(\frac{M_e^* c^2}{16\pi e^2 n_0}\right)^{1/2}, \tag{5.3.6}$$

c is the velocity of light, e is the electron charge, M_e^* is the bipolaron mass, n_0 is the concentration of TI bipolarons in Bose condensate, $n_0 = N_0/V$, and N_0 is the number of TI bipolarons in the condensate:

$$\frac{N_0}{N} = 1 - \frac{T^{3/2}}{C_{bp}} F_{3/2}(\omega_0/T), \tag{5.3.7}$$

N is the total number of bipolarons, $C_{bp} = \left(\frac{n^{2/3} 2\pi\hbar^2}{M_e \omega^*}\right)^{3/2}$, $\tilde{T} = T/\omega^*$, ω^* is an arbitrary energy scale factor, $n = N/V$.

As is noted in Lakhno (2020a), the mass of a TI bipolaron does not differ too much from 2 m, where m is the mass of a band electron which depends on ω_0 only slightly. Therefore, we will believe that the whole dependence on ω_0 is determined by the concentration of TI bipolarons in the condensate n_0 involved in (5.3.6) which is related with ω_0 by relation (5.3.7). As a result, from (5.3.5) to (5.3.7), we express the isotopic coefficient β as

$$\beta = -\frac{1}{2}\frac{\tilde{\omega}_0 \tilde{T}^{1/2} N}{C_{bp} N_0} Li_{1/2}\left(e^{-\omega_0/T}\right),$$

$$Li_{1/2}(Z) = \frac{1}{\sqrt{\pi}} \int_0^\infty \frac{1}{\sqrt{t}} \frac{dt}{Z^{-1} e^t - 1}. \tag{5.3.8}$$

It follows from (5.3.8) that in the limit of low temperatures when $\omega_0/T \gg 1$, $Li_{1/2}(e^{-\omega_0/T}) = e^{-\omega_0/T}$:

$$\beta = -\frac{1}{2}\frac{\tilde{\omega}_0 \tilde{T}^{1/2}}{C_{bp}} e^{-\omega_0/T}, \tag{5.3.9}$$

that is the isotopic coefficient is exponentially small.

In the case of $\omega_0/T \ll 1$, $Li_{1/2}(e^{-\omega_0/T}) = \sqrt{\pi T/\omega_0}$, the isotopic coefficient is equal to:

$$\beta = -\frac{\sqrt{\pi}}{2}\frac{\tilde{\omega}_0^{1/2} \tilde{T} N}{N_0 C_{bp}} = -\frac{\sqrt{\pi}}{2}\frac{\tilde{\omega}_0^{3/2} N}{N_0 C_{bp}}\left(\frac{T}{\omega_0}\right). \tag{5.3.10}$$

It should be noted that, as distinct from the coefficient α, which is positive, the coefficient β for London penetration depth is negative, which is in agreement with experiment. The fact that in the limit of low temperatures (5.3.9) the isotopic coefficient β, caused by EPI is negligible is in agreement with the BCS (Bill et al., 1998a, 1998b). This in particular implies that in the limit of weak doping (when $T_c \to 0$) the isotopic coefficient for the London penetration depth β (as distinct from the isotopic coefficient α for

T_c (which will be large in this case)) will be very small. The experiment, however, demonstrates that the value of β in the case of optimal doping (when T_c is maximum) in the limit of low temperatures can reach large values. Thus, for example, in the optimally doped HTSC $YBa_2Cu_3O_7$ at $T = 0$ the value of β is equal to $|\beta(0)| \approx 0.2$ (Khasanov et al., 2004). It follows that the main contribution in this case can be made by nonadiabatic mechanism or nonphonon mechanisms (Bill et al., 1998a, 1998b). As an example of HTSC, in which the contribution of nonadiabaticity and nonphonon mechanisms is probably small we can present a slightly overdoped $La_{2-x}Sr_xCu_{1-y}Zn_yO_4$, for which at $T = 0$ the isotopic coefficient β vanishes in accordance with the theory developed (Tallon et al., 2005). Experimental results on the isotope dependence of the penetration depth in $La_{2-x}Sr_xCu_{1-y}Zn_yO_4$ are often explained by the isotope dependence of the effective mass of current carriers:

$$\Delta M_{bp}/M_{bp} = 2\Delta\lambda/\lambda + \Delta n_0/n_0. \tag{5.3.11}$$

Thus, it is stated that we can neglect a change in n_0 due to the isotope effect, and the entire effect can be attributed to the change in effective mass M_{bp} (Zhao et al., 1997). This statement could be true if the BCS theory be applicable, in which the quantity n_0 coincides with the total number of electrons in the normal phase. In the BCS theory, however, the effective mass of current carriers does not depend on the masses of the lattice atoms. As is shown in Lakhno (2018), n_0 represents only a small portion of the number of normal electrons. This is confirmed by the experiments performed by Božović et al. (2016). As is shown above, the use of this fact enables one to explain the isotope effect for the London penetration depth by the isotope dependence of the quantity n_0.

At high temperatures, on the contrary, the main contribution near T_c can be made by the phonon mechanism determined by (5.3.10) according to Bill et al. (1998a, 1998b), for the value $\alpha = 0.025$, observed in $YBa_2Cu_3O_{7-\delta}$ β was obtained to be $\beta \sim -0.6$ for $T/T_c \sim 0.95$).

The results obtained enable one to explain the observed peculiarities of the behavior of the isotopic coefficient α for T_c in high-temperature superconductors, in particular, its small value for optimal doping and its large value for weak doping, on the basis of only EPI. More complicated is the question of the isotopic coefficient for the London penetration depth λ and its temperature dependence.

The EPI explains the large values for optimally doped HTSC materials only near the SC transition temperature T_c. In the case of low temperatures, the theory explains the negligible values of $\beta(0)$ in such HTSC as $La_{2-x}Sr_xCu_{1-y}Zn_yO_4$ and does not explain large values in other HTSC compounds. In Bill et al. (1998a, 1998b), this discrepancy with the theory developed and the BCS is explained by the fact that in many HTSC materials at low temperatures, the main role is played by nondiabatic effects, leading to large values of $\beta(0)$.

5.4 Conclusive remarks

In the theory presented, as in the BCS, the momentum of the bipolaron mass center **P** (accordingly **G** in a magnetic field) is equal to zero. According to this theory, the SC state is a homogeneous bipolaron BEC. The theory can be generalized to the case of a moving BEC with $\mathbf{P} \neq 0$, which remains homogeneous when moving. In this case, some interesting peculiarities arise (Chapter 6). Presently, a wide discussion is devoted to the possibility of the formation of an inhomogeneous BEC in the form of the so-called pair density waves (PDW) which destroy the translation invariance (Lee, 2014; Seo et al., 2008; Berg et al., 2009; Agterberg and Tsunetsugu, 2008; Zelli et al., 2012; Chen et al., 2004; see Chapter 7). However, here the situation is different from the problem of polarons and bipolarons with broken or actual TI symmetry. The scenario of SC with PDW including the presence of charged density waves (CDW) or spin density waves (SDW) (Pépin et al., 2014; Freire et al., 2015; Wang et al., 2015a, 2015b) is provided by the discreteness of the crystal which is not taken into account in the continuum model of EPI. The problem of the competition between the CDW mechanism of SC and the bipolaron one is considered, for example, in Grzybowski and Micnas (2007) for a SRP in squeezed vacuum.

Modulation of BEC density for wave vector corresponding to nesting leads to the appearance of a gap in the spectrum which in many works is identified with a SC gap (Lee, 2014; Seo et al., 2008; Berg et al., 2009; Agterberg and Tsunetsugu, 2008; Zelli et al., 2012; Chen et al., 2004; Pépin et al., 2014; Freire et al., 2015; Wang et al., 2015a, 2015b). In this case, the TI bipolaron gap ω_k, being universal, would have the properties of a pseudogap manifesting itself as a low-energy thin structure in the conductance spectrum of optimally doped SC (Maggio-Aprile et al., 1995; Pan et al., 2000; Hoogenboom et al., 2000).

In the approach considered, we actually did not use any specificity of the mechanism of the electron or hole pairing. For example, both in the Hubbard model and in the $t-J$ model, in describing copper oxide HTSC the same holes take part in the formation of antiferromagnetic fluctuations and pairing caused by an exchange by these fluctuations. If an interaction of holes with magnetic fluctuations leads to the formation of TI magnetopolarons having the spectrum $\omega_0(k)$, then this spectrum is also the spectrum of magnons renormalized by their interaction with holes (bound magnons). For this reason, the statement that the RVB superconductor is just a limiting case of the BCS SC with strong interaction becomes justified (Kivelson and Rokhsar, 1990; in this case, the role of polarons and bipolarons belongs to holons and biholons).

Evidently, *d*-symmetry is specificity of cuprate HTSC and is not a precondition of the existence of HTSC. For example, sulfide H_2S, demonstrates a record value of the transition temperature: $T_C = 203$ K (under high pressure, Drozdov et al., 2015), does not have a magnetic order, but EPI is strong in it. Still greater value of T_C under high pressure has recently been obtained in the substance LaH_{10} with $T_C = 260$ K (Somayazulu et al., 2019), where the EPI is also strong and a magnetic order is lacking. Finally, the room transition temperatures (about 15 °C) were obtained in composition on the basis of H_2S and CH_4 (under the pressure of about of 1.4 million atmosphere) in Snider et al. (2020).

Nevertheless, the mechanism of pairing is still unclear. If it is provided by an interaction of current carriers with magnetic fluctuations, then, in the approach considered, the particles which bind electrons into pairs will be magnons rather than phonons. In passing on from the pseudophase to the normal one, this binding mode disappears which leads to the decay of a bipolaron into two individual polarons with the emission of a phonon (magnon).

In the pseudogap phase, there may be a lot of different gaps caused by the presence of phonons, magnons, plasmons, and other types of elementary excitations. In this case the SC gap will be determined by the type of elementary excitations whose interaction with the current carriers is the strongest.

From the viewpoint of the TI bipolaron theory, a possible resultant picture of HTSC looks as follows.

According to the above consideration, the foundation of the microscopic theory is provided by the TI bipolaron EPI mechanism. It follows from the theory that in order to reach high T_C one should primarily enhance the concentration of TI bipolarons. In oxide ceramics, this is reached by the presence of antiferromagnetic order and stripes in them.

Playing the role of microscopic domain walls, the stripes, having a ferromagnetic order, attract electrons. Because of the exchange interaction, the energy of electrons in the stripes is lower than that in the rest of the template (analog of ferrons by Nagaev (1979) with regard to the contribution of polaron (Lakhno and Nagaev, 1976) and magnetostriction effects (Lakhno and Nagaev, 1978) into their formation), accordingly, the concentration of electrons there is rather high. To restore a charge equilibrium TI bipolarons flow from the stripe regions to the template thus enhancing the concentration of TI bipolarons in it and, on the whole T_C of the sample. This redistribution gives rise to a PDW (elevated concentration of bipolarons in the template and reduced concentration in the stripes) and CDW (elevated concentration of electrons in the stripes and reduced concentration in the template).

The mechanism described enables one to construct purposively SC materials which could work at room T_C. As it was pointed out in Lakhno (2019b), to do so one can use inhomogeneous doping making the periphery of a HTSC cable doped with ferromagnetic impurities which could attract electrons from the core of the cable. As a result, one can take a cable with enhanced concentration of TI bipolarons on its axis and, as a consequence, high T_C.

6 Moving bipolaron Bose condensate

In the previous chapters, we dealt with the case of resting Bose condensate of TI bipolarons. In this chapter, this case is generalized to the case of a moving Bose condensate.

6.1 Introduction

The translation-invariant bipolaron theory suggests that in HTSC materials bipolarons represent almost an ideal Bose gas or Bose liquid, if we are talking about their Bose condensate, which can move. An example of a nonideal Bose fluid is a quantum fluid such as ^{4}He. The difference between an ideal and a nonideal Bose fluid is in the spectrum of their elementary excitations. In a nonideal Bose liquid, in the case of small momenta, there is a phonon branch $\varepsilon = sk$, where s is the sound velocity: $s = \partial P/\partial \rho$, where P is the pressure, ρ is the liquid density. In the case of large k, the spectrum becomes quadratic $\varepsilon = \varepsilon(k_0) + (k - k_0)^2/2M$ which is called roton.

In an ideal Bose liquid, the pressure does not depend on the density, that is, $s = 0$, and the linear part is lacking in the spectrum. Such a spectrum (with $k_0 = 0$) was considered in the original work by Landau (1941). As is shown above, such a spectrum corresponds to a gas of TI bipolarons.

The fundamental difference between the TI bipolaron Bose gas and the superfluid Bose liquid in helium is that the elementary quasiparticles of the TI bipolaron gas are charged, while the elementary excitations in helium are neutral.

Thus, superconductivity is a superfluidity of a charged Bose gas. This point of view leads to the possibility of using the results of the theory of superfluidity as applied to superconductivity in the macroscopic description of the latter.

6.2 Moving Bose condensate of TI bipolarons

In Landau (1941), he developed a two-fluid theory of superfluid helium II as an alternative to the theories by London (1938) and Tisza (1938), which related this phenomenon to Bose–Einstein condensation. These two extreme points of view were reconciled by Bogolyubov, who, using a weakly imperfect Bose gas as an example, reproduced the phonon–roton spectrum of the two-fluid Landau model and showed that the superfluid component in this case is a condensate of Bose particles (Bogolyubov, 1947). This, however, did not happen in the case of superconductivity. As was pointed out by BCS (Bardeen et al., 1957), Cooper pairs in ordinary metals are unsuitable for the role of Bose particles due to their enormous overlapping. The idea that superfluidity and superconductivity are related phenomena was strengthened

https://doi.org/10.1515/9783110786668-006

only after the discovery of HTSC, when it was found that paired states in these materials have a small correlation length.

Presently, there is a large number of candidates for the role of the fundamental Bose particle responsible for HTSC. There are, however, basically two points of view competing. According to one, the pairing mechanism is caused, as in the case of the BCS, by the electron-phonon interaction (Bardeen et al., 1957). According to the other, the pairing of current carriers is caused by magnetic fluctuations (Scalapino, 2012).

Without going into details of the argumentation in favor of one point of view or another, let us choose the spectrum of the TI bipolaron as the spectrum of the fundamental boson responsible for HTSC. The theory of HTSC based on the theory of TI bipolarons was previously constructed by us using its spectrum of excited states:

$$\varepsilon(\mathbf{k}) = \omega_0 + \mathbf{k}^2/2M, \quad \mathbf{k} \neq 0, \tag{6.2.1}$$

where $M = 2m$, m is the effective mass of a band electron which exactly coincides with roton spectrum in the superfluidity theory by Landau (1941).

At $T = 0$ all the bipolarons are in the condensed state. If Bose condensate moves relative to the crystal lattice of the sample, then the total momentum of the Bose condensate relative to the lattice will be equal to $\mathbf{P}$:

$$\mathbf{P} = \sum \mathbf{k}\overline{m}(\mathbf{k}), \tag{6.2.2}$$

where $\overline{m}(\mathbf{k})$ is Bose function of the distribution of TI bipolarons. In a condensed state each TI bipolaron has one and the same momentum: $\mathbf{k}_u = M_{bp}\mathbf{u}$, where $\mathbf{u}$ is the velocity of a TI bipolaron in a condensate (i.e., the velocity of a Bose condensate), M_{bp} is the mass of a TI bipolaron. Accordingly, the function of distribution $\overline{m}(\mathbf{k})$ in this case will be:

$$\overline{m}(\mathbf{k}) = N_o\Delta(\mathbf{k} - \mathbf{k}_u), \tag{6.2.3}$$

where N_0 is the number of bipolarons in a condensate which at $T = 0$ is equal to the total number of TI bipolarons N. Hence, the total momentum of Bose condensate at $T = 0$ will obviously be equal to: $\mathbf{P} = N_0 M_{bp}\mathbf{u} = NM_{bp}\mathbf{u}$.

Now let us consider the case of nonzero temperature $T < T_c$, where T_c is the temperature of a superconducting transition. In this case some bipolarons are in an excited state. Being in an excited state, a bipolaron can interact with other excitations and defects of a crystal. As a result of such interaction a gas of excited states, being in equilibrium with the lattice, as a whole rests relative the lattice. At the same time, the gas of excitations cannot put stay the condensate part, since it cannot exchange momentum with it (Lifshitz and Pitaevskii, 1980). As a result, the distribution function of all TI bipolarons will have the form:

$$\overline{m}(\mathbf{k}) = N_0\Delta(\mathbf{k}) + \left[\exp\left(\frac{\varepsilon(\mathbf{k}) - \mathbf{ku}}{T}\right) - 1\right]^{-1}(1 - \Delta(\mathbf{k})), \tag{6.2.4}$$

where $\varepsilon(\mathbf{k})$ is the spectrum of excited states of a TI bipolaron, $\mathbf{u}$ is the velocity of a Bose condensate relative the crystal lattice. The expression takes into account that if the excitation spectrum of TI bipolarons in a system of a resting Bose condensate has the form $\varepsilon(\mathbf{k})$ then in the system where the excitation gas rests (i.e., in the system related with the lattice), the relevant spectrum will be $\varepsilon(\mathbf{k}) - \mathbf{ku}$.

Substitution of (6.2.1) and (6.2.4) into (6.2.2) yields the following value of the total momentum $\mathbf{P}'$ of excitations in the system of a resting condensate:

$$\mathbf{P}' = -M_{bp}\mathbf{u}N', \tag{6.2.5}$$

$$N'/V = \left(MT/2\pi\hbar^2\right)^{3/2} F_{3/2}(\tilde{\omega}/T), \tag{6.2.6}$$

$$F_{3/2}(\alpha) = \frac{2}{\sqrt{\pi}}\int_0^\infty \frac{x^{1/2}}{e^{x+\alpha} - 1}dx, \tag{6.2.7}$$

$$\tilde{\omega}_0 = \omega_0 - Mu^2/2, \tag{6.2.8}$$

where V is the crystal volume, $N' = N - N_0$. Hence, the total momentum of TI bipolarons in a laboratory frame of reference, that is, in the system related with the crystal lattice will be equal to:

$$\mathbf{P} = (N - N')M_{bp}\mathbf{u} = N_0 M_{bp}\mathbf{u}. \tag{6.2.9}$$

It follows from (6.2.6) to (6.2.8) that there exists a limit velocity of the motion of a Bose condensate u_c:

$$u < u_c, \qquad u_c = \sqrt{2\omega_0/M}. \tag{6.2.10}$$

Condition (6.2.10) exactly corresponds to Landau's criterion of superfluidity:

$$u < \varepsilon(\mathbf{k})/k. \tag{6.2.11}$$

It follows from (6.2.6) to (6.2.8) that the expression for the temperature of a superconducting transition (which is derived from (6.2.6) for $N' = N$) will be the same as that in Lakhno (2018, 2019b, 2020a) (Chapter 4) with ω_0 replaced by $\tilde{\omega}_0$, which is determined by (6.2.8). Hence, the temperature of a superconducting transition depends on the velocity of the Bose condensate motion and reaches its maximum value for $u = 0$. As the condensate velocity increases, T_c decreases and reaches its minimum: $T_c = 3.31\hbar^2 n^{2/3}/M$, $n = N/V$ is the concentration of TI bipolarons for $u = u_c$.

What will happen if the velocity of the motion of a Bose condensate exceeds a certain critical value, that is, if the inequality $u > u_c$ holds? The integral in (6.2.7) in this case does not exist and a stationary motion turns out to be impossible. For $\mathbf{P} > \mathbf{P}_c = N_0 M_{bp} u_c$ the momentum from the excitations begins to be transmitted to the condensate, hindering it till the velocity of the condensate becomes equal to u_c.

6.3 Bose condensate in a magnetic field

Let us consider the case when a sample is placed in a magnetic field. In view of Meissner effect, the magnetic field sets in motion the Bose condensate in the near-surface layer of the sample whose thickness is of the order of the London penetration depth. Since the magnetic field does not penetrate into the sample, the bulk of the Bose condensate in the sample will be immobile. The total magnetic field acting on individual bipolarons includes both the external field and the field created by the moving Bose condensate itself. Let us consider the contribution of the field induced by the moving Bose condensate (which will be denoted by **B**) into the current.

According to Lakhno (2019b, 2020a) (Chapter 3), the excitation spectrum of the Bose condensate of TI bipolarons in a magnetic field of intensity **B** has the form:

$$\varepsilon_B(k) = \omega_0 + \frac{k^2}{2M} + \frac{\eta}{M}(\mathbf{Bk}) - \mathbf{ku}. \tag{6.3.1}$$

It follows from (6.3.1) that in the case of a magnetic field all the above formulae do not change their form if the velocity $\mathbf{u}$ in them is replaced by $\tilde{\mathbf{u}} = \mathbf{u} - \eta\mathbf{B}/M$. The field induced by a Bose condensate moving with the velocity $\mathbf{u}(r)$ at each point of the sample $\mathbf{r}$ will always be perpendicular to $\mathbf{u}(r)$. Hence, in a magnetic field, instead of the quantity $\tilde{\omega}_0$, determined by (6.2.8), we will get the quantity $\tilde{\omega}_{0,B}$, equal to:

$$\tilde{\omega}_{0,B} = \omega_0\left(1 - \frac{u^2}{u_c^2} - \frac{B^2}{B_{\max}^2}\right), \tag{6.3.2}$$

where $B_{\max} = \sqrt{2M\omega_0/\eta}$. The quantity $B_{\max}$ has the meaning of the maximum value of the magnetic field (at $T = 0$), for which the superconductivity takes place as $\mathbf{u} = 0$ (first critical field).

According to the results obtained, the moving near-surface Bose condensate should have a much lower condensation temperature $T_c(\mathbf{u})$, than a condensate at rest in the sample, whose condensation temperature is $T_c(\mathbf{u} = 0, \mathbf{B} = 0) > T_c(\mathbf{u} \neq 0, \mathbf{B} \neq 0)$, as it follows from (6.2.6) to (6.2.8) and (6.3.2).

It follows from (6.3.2) that for the value of $\tilde{\omega}_{0,B}$ to be positive, the condition:

$$u < u_c\sqrt{1 - B^2/B_{\max}^2}. \tag{6.3.3}$$

should be fulfilled. Hence, the dependence of the value of the critical current in the superconductor on the magnetic field B, has the form:

$$j_c(B) = j_c(0)\sqrt{1 - B^2/B^2_{\text{max}}}, \qquad j_c(0) = 2en_0u_c, \tag{6.3.4}$$

where n_0 is the concentration of TI bipolarons in a Bose condensate.

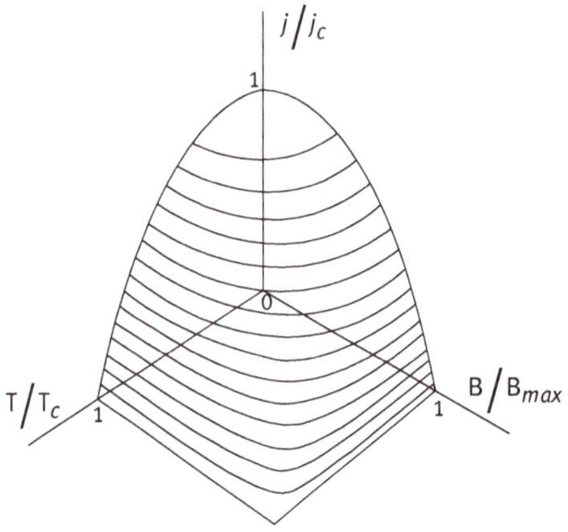

Fig. 6.1: Dependence of the critical current on the magnetic field and temperature.

With regard to the temperature dependence $n_0(T)$ (Fig. 4.3) we get from (6.3.4) an ordinary dependence for the critical value of the current density $j_c(B,T)$ as a function of the temperature and magnetic field B, presented in Fig. 6.1.

Expression (6.3.2) was obtained on the assumption that the quantity **B** is a field induced by a moving Bose condensate. However, according to the Silsbee rule (Silsbee, 1917), the action of the field **B** is independent of whether it is induced by a current running in a superconductor (i.e., a moving Bose condensate) or generated by an external source. Hence, Fig. 6.1 can be considered as a dependence of the critical current on the external magnetic field and temperature.

6.4 Little–Parks effect

Little–Parks effect (Little and Parks, 1962) is concerned with the fact that quantizing of a magnetic field in a multiply connected SC is caused by relevant quantizing of a current running around each hole in a sample. Quantizing of a current, leads to oscillations of the velocity of the superconducting current carriers. Since the temperature of a superconducting transition depends on the velocity of a Bose condensate, its oscillations lead to relevant oscillations of the temperature of a superconducting transition. The study of this phenomenon has recently attracted much attention of researchers due to the possibility to use it for enhancing the critical temperature in SC (Sochnikov et al., 2010; de Gennes, 1981; Staley and Liu, 2012; Liu et al., 2001).

Since at present a generally accepted theory of high-temperature superconductivity is lacking, a generally accepted explanation of the Little–Parks effect in HTSC is also lacking. The aim of this section is to develop the theory of the Little–Parks effect on the basis of the translation invariant bipolaron theory (Lakhno, 2019c).

With the use of (6.3.1) we can get an expression similar to (6.2.5)–(6.2.8), which instead of $\tilde{\omega}_0$, determined by (6.2.8), will involve $\tilde{\omega}_{0,B}$, determined by (6.3.2). From the general relations for quantizing the fluxoid Φ' we get: $\Phi' = n\Phi_0$, $n = 1, 2 \ldots$, $\Phi_0 = \pi \hbar c/e$, which in the case of a bipolaron Bose condensate has the form:

$$\Phi' - \Phi = \frac{c}{2e}\oint M_{bp}\mathbf{u}d\mathbf{R}, \quad \Phi = \oint \mathbf{A}d\mathbf{R}, \tag{6.4.1}$$

where Φ is a magnetic flux passing through the contour of integration $\mathbf{R}$:$\Phi = \int \mathbf{B}d\mathbf{S}$. In the case of a thin-walled cylinder corresponding to the conditions of the Little–Parks experiment, the expression for the velocity of a TI bipolaron Bose condensate is written as follows:

$$u = \frac{\hbar}{M_{bp}R}\left(n - \frac{\Phi}{\Phi_0}\right), \tag{6.4.2}$$

where R is the radius of the cylinder. Equation (6.4.2) describes oscillations of the velocity of a TI bipolaron Bose condensate (Fig. 6.2) which leads to oscillations of the temperature of the superconducting transition T_c (Tinkham, 1975). Using (6.2.5)–(6.2.7) and (6.3.2) and assuming that $\tilde{\omega} = \tilde{\omega}_0 + \Delta\tilde{\omega}$, where $\tilde{\omega}_0 = \omega_0\left(1 - \Phi^2/\Phi_c^2\right)$, $\Delta\tilde{\omega} = -\omega_0 u^2/u_c^2$, where Φ_c is the critical magnetic flux corresponding to the critical magnetic field B_c, for small deviations $\Delta\tilde{\omega}$ for deviations of the critical temperature from that determined by (6.2.6) we get:

$$\frac{\Delta T_c}{T_c} = \frac{\xi^2}{R^2}\left(n - \frac{\Phi}{\Phi_0}\right)^2, \tag{6.4.3}$$

$$\xi^2 = -\frac{1}{3(2\pi)^{3/2}}\frac{M^2}{M_{bp}^2}\frac{1}{n_{bp}^{2/3}}\left(\frac{MT_c}{n_{bp}^{2/3}\hbar^2}\right)^{1/2} Li_{1/2}(e^{-\alpha}), \tag{6.4.4}$$

$$Li_{1/2}(e^{-\alpha}) = \frac{1}{\sqrt{\pi}}\int_0^\infty \frac{1}{\sqrt{t}}\frac{dt}{e^{t+\alpha} - 1}, \tag{6.4.5}$$

$$\alpha = \frac{\omega_0}{T_c}\left(1 - \frac{\Phi^2}{\Phi_c^2}\right), \tag{6.4.6}$$

where n_{bp} is the concentration of TI bipolarons. Formula (6.4.3) determines the amplitude of T_c oscillations as the magnetic flux Φ changes. Despite the oscillating behavior of ΔT_c depending on the magnetic flux, expression (6.3.2) assumes that there is a maximally possible value of the magnetic flux $\Phi_{\max} = \Phi_c = \pi R^2 B_c$, corresponding to the maximum value of the magnetic field for which a Bose condensate with $T_c(\Phi = \Phi_c, u = 0) = 3,31\hbar^2 n_{bp}^{2/3}/M$ can exist.

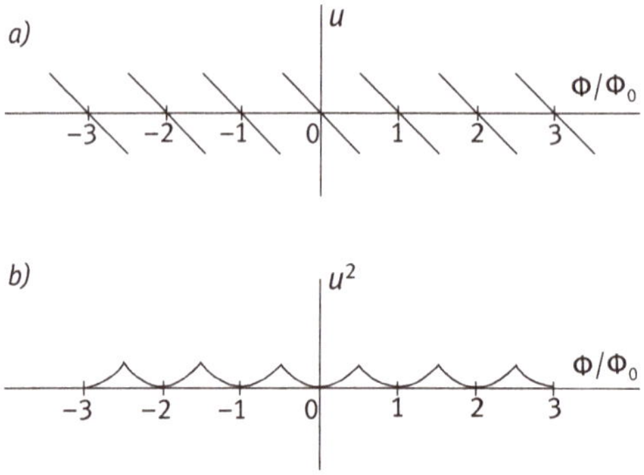

Fig. 6.2: (a) Dependence of the quantity u on the magnetic flux passing through the cylinder bore, (b) dependence of the quantity u^2 and, accordingly, $\Delta T_c \sim u^2$ on the magnetic flux. (Tinkham, 1975).

For low critical temperatures: $\alpha \gg 1$, $Li_{1/2}(e^{-\alpha}) \cong e^{-\alpha}$, while for high ones: $\alpha \ll 1$, $Li_{1/2}(e^{-\alpha}) \approx \sqrt{\pi/\alpha}$. With the use of expressions (6.4.3) and (6.4.4) for $\alpha \cong 1$, $n_{bp} = 10^{19}$ cm^{-3} (Lakhno, 2018), $m = m_0$, $M \approx M_{bp}$, $R = 100$nm for $\Delta T_c/T_c$ we get: $\Delta T_c/T_c \approx 10^{-4}$, which is close to the estimate obtained on the basis of the BCS (Tinkham, 1975). It also follows from (6.4.3) and (6.4.4) that for $\omega_0/T_c \ll 1$ (accordingly, $\alpha \ll 1$) the relation: $\Delta T_c/T_c \approx 10^{-4}(T_c/\omega_0)^{1/2}$ can be much greater than that obtained on the basis of the BCS theory.

The Little–Parks effect is usually used to check the fundamental positions of the superconductivity theory, in particular, to prove pairing of current carriers in superconductors. The results obtained can also be used for finding soft phonon modes for which $\omega_0 \to 0$ (see Kimura et al., 2005), which usually accompany structural instability and phase transitions in HTSC materials. In this case one can expect anomalous values of $\Delta T_c/T_c$ oscillations.

Presently, the main research methods of soft modes investigation are inelastic X-ray scattering and inelastic neutron scattering (Wakimoto et al., 2004; Reznik, 2010). The accuracy of these methods is limited to energies of a few millielectronvolts. As a result, the estimate of the value of $(T_c/\omega_0)^{1/2}$ will give a small coefficient: <10. For real values of T_c equal to several tens of degrees this can lead to difficulties in interpreting the data obtained on the basis of the Little–Parks effect.

This conclusion is confirmed by observations of anomalous softening of phonon modes in the family of $La_{2-x}Sr_xCuO_4$ compounds (Kimura et al., 2005) for values of x close to those for which the ratio $\Delta T_c/T_c$ can be anomalously large (Sochnikov et al., 2010), which were associated in (Sochnikov et al., 2010) with the possibility of vortex formation and their interaction with an oscillating current.

The results obtained here are based on the idea of Bose condensation of TI bipolarons with a small correlation length (≈ 1 nm). This is true only in the case of a strong EPI, when the EPI constant is anomalously large. The condition $\alpha_{ph} \approx \infty$ is fulfilled only as $\omega_0 \to 0$, since $\alpha_{ph} \sim \omega_0^{-1/2}$. Thus, bipolarons formed by interaction with such phonon modes will have the lowest energy compared to those formed by other branches of the phonon spectrum and, hence, it is these phonons that will form the Bose condensate and lead to anomalies in the Little–Parks effect (see Chapter 7).

Above, we considered the thermodynamic characteristics of a Bose condensate with an isotropic phonon spectrum. In fact, in most HTSC materials, the SC gap depends on the wave vector. For example, in YBCO with optimal doping, the dependence of the gap on $\mathbf{k}$ has the form:

$$\omega_0(\mathbf{k}) = \Delta_0 \left|\cos k_x a - \cos k_y a\right| + \omega_0, \tag{6.4.7}$$

where $\omega_0(\mathbf{k})$, according to the above, is the phonon frequency which depends on $\mathbf{k}$. In TI-bipolaron theory, the first term on the right-hand side of (6.4.7) corresponds to the contribution of the d-type wave, and the second term to the contribution of the s-type wave.

Hence, for $\omega_0 \geq \Delta_0 Ma^2 T/\hbar^2$ the main contribution into the integral (6.2.7) will be made by s-wave (see Section 4.5). In this case the results obtained above will remain unchanged if we replace ω by ω_0. Thus, for example in the case of YBCO the values of ω_0/Δ_0 is ≈ 0.15 (Smilde et al., 2005; Kirtley et al., 2006), that is, ω_0 is sufficiently small.

Expressions (6.4.3)–(6.4.6) also show that in the general case, when softening of the phonon modes is small to achieve the maximum amplitudes of T_c oscillations in HTSC materials, it is necessary to use the maximum values of the magnetic flux $\Phi \to \Phi_c$. In this case, an arbitrarily small deviation of the Bose condensate velocity from its equilibrium value $\Phi \to \Phi_c$ will lead to finite changes of T_c. This effect can be used for construction of new types of magnetic field sensors and switchers between superconducting and normal states by means of small changes in magnetic field.

7 Translation-invariant bipolarons and charge density waves

A relation is established between the theories of superconductivity based on the concept of charge density waves (CDW) and the translation-invariant bipolaron theory. It is shown that CDWs are formed in the pseudogap phase from bipolaron states due to Kohn anomaly, forming a pair density wave (PDW) for wave vectors corresponding to nesting. Formed in the pseudogap phase, CDWs coexist with superconductivity and their amplitude decreases as a TI bipolaron Bose condensate is formed and vanishes at zero temperature.

7.1 Translation-invariant bipolarons and charge density waves in high-temperature superconductors

Presently, there is no consensus on the microscopic nature of high-temperature superconductivity (HTSC). At the same time, there are phenomenological models such as the Ginzburg–Landau model (Larkin and Varlamova, 2005), the CDW or PDW model, and spin density waves which enable describing numerous HTSC experiments (Grüner, 1994). These models do not say anything about the nature of the paired states participating in the SC. In this monograph, paired states mean translation-invariant (TI) bipolaron states formed by a strong electron-phonon interaction, similar to Cooper pairs. According to the above results, TI bipolarons are plane waves with a small correlation length which can form a Bose–Einstein condensate with a high transition temperature, possessing SC properties. The relation between the Bardeen–Cooper–Schrieffer theory (Bardeen et al., 1957) (BCS) and the Ginzburg–Landau theory was established in Gor'kov (1959). The aim of this chapter is to establish a relation between the TI bipolaron theory of SC and CDW (PDW).

7.1.1 General relations for the spectrum of a moving TI bipolaron

According to Chapter 4, TI bipolarons are formed at a temperature T^* much higher than the temperature of a SC transition T_c. For $T_c < T < T^*$ and in the absence of a Fermi surface with a sharp boundary, an ensemble of TI bipolarons would be an ideal gas, whose particles would have a spectrum determined by the dispersion equation (4.2.19) of Chapter 4:

$$1 = \frac{2}{3}\sum_k \frac{k^2|f_k|^2\omega_k}{s-\omega_k^2}, \tag{7.1.1}$$

$$\omega_k = \omega_0(\mathbf{k},\mathbf{P}) - \frac{\mathbf{kP}}{M} + \frac{k^2}{2M} - \frac{\mathbf{k}}{M}\sum_{k'}\mathbf{k}'|f_{k'}|^2,$$

https://doi.org/10.1515/9783110786668-007

$f_k = f_k(\mathbf{k}, \mathbf{P})$ are parameters determining the ground-state energy of a TI bipolaron, $E_{bp}(\mathbf{P})$, $\mathbf{P}$ is the total momentum of a TI bipolaron, $M = 2m$, where m is the mass of a band electron (hole), $\omega_0(\mathbf{k}, \mathbf{P})$ is the phonon frequency in an electron gas surrounding TI bipolarons (see Appendix I).

A wave function of a TI bipolaron with the wave vector $\mathbf{P}$ will have the form:

$$|\Psi(\mathbf{P})\rangle_{bp} = e^{i\mathbf{P}\mathbf{R}}|\Psi(0)\rangle_{bp}, \tag{7.1.2}$$

where $\mathbf{R}$ is coordinates of the center of mass of a bipolaron. An explicit form of the wave function $|\Psi(0)\rangle_{bp}$ with zero momentum is given in (2.11.13) Chapter 2. An expression for $\mathbf{P}$ can be obtained by calculating the mathematical expectation of the total momentum operator $\mathbf{P}$:

$$\mathbf{P} = \langle\Psi(\mathbf{P})|\widehat{\mathbf{P}}|\Psi(\mathbf{P})\rangle = M\mathbf{u} + \sum_k \mathbf{k}|f_k|^2, \tag{7.1.3}$$

where $\mathbf{u}$ is the velocity of a TI bipolaron. Assuming $\mathbf{P} = M_{bp}\mathbf{u}$, where M_{bp} is the bipolaron mass, we express M_{bp} with the use of (7.1.3) as

$$M_{bp} = \frac{M}{1-\eta}, \quad \eta = \frac{\mathbf{P}}{\mathbf{P}^2}\sum_k \mathbf{k}|f_k|^2. \tag{7.1.4}$$

With the use of (7.1.4) the expression for ω_k from (7.1.1) can be rewritten as

$$\omega_k = \omega_0(\mathbf{k}, \mathbf{P}) + \frac{k^2}{2M} - \frac{\mathbf{k}\mathbf{P}}{M_{bp}}. \tag{7.1.5}$$

It follows from (7.1.4) that in the case of weak and intermediate coupling (when TI bipolaron states are metastable for $\mathbf{P} = 0$), the exact form of $f_k(\mathbf{k}, \mathbf{P})$ is known and the expression for the effective mass of a TI bipolaron will have a simple form: $M_{bp} = M(1 + \alpha/6)$, where α is a constant of electron–phonon coupling, that is, the mass M_{bp} is equal to the sum of masses of individual polarons. When α is large, good approximations for f_k are available only for $\mathbf{P} = 0$. For this reason, the calculation of the effective mass of a strong-coupling TI bipolaron is rather difficult.

7.1.2 TI bipolarons and charge density waves

If there is a Fermi surface with a sharp boundary, the TI bipolaron gas will have some peculiarities. Thus, if on this surface, there are sufficiently large fragments that can be combined by transferring one of them to vector $\mathbf{P}$, then, given a sufficient size of these fragments, the coupling between them will be strong enough, which will lead to Peierls deformation of the lattice in the direction of such nesting. The loss in the energy associated with lattice deformation will be compensated by a gain in the energy of the bipolaron gas, which forms a CDW with the wave vector $\mathbf{P} = \mathbf{P}_{CDW}$. The availability of a gain in the TI bipolaron energy follows from eq. (7.1.1), solution of which leads to the spectrum of a TI bipolaron:

$$E_k(\mathbf{P}) = \begin{cases} E_{bp}(\mathbf{P}), & k = 0 \\ E_{bp}(\mathbf{P}) + \omega_0(\mathbf{P}, \mathbf{k}) + \frac{k^2}{2M} - \frac{\mathbf{kP}}{M_{bp}}, & k \neq 0. \end{cases} \tag{7.1.6}$$

that is, $E_k = E_{bp} + \omega_k$, and $v_k = \omega_k$ for $k \neq 0$.

A gain in the energy is caused by the so-called Kohn anomaly (Kohn, 1959), which implies that if $\mathbf{P} = \mathbf{P}_{CDW}$ softening of a phonon frequency takes place (in 1D metal $\mathbf{P}_{CDW} = 2k_F$, $\omega_0(2k_F) = 0$, where k_F is the Fermi momentum) $\omega_0(\mathbf{P}_{CDW}, k) = \omega_{CDW}$ and, as a result, a great decrease in the energy E_{bp} in view of a sharp increase of the electron-phonon coupling constant $\alpha \sim 1/\omega_{CDW}^{1/2}$ (Appendix I). As $\omega_{CDW} \to 0$, $\alpha_{CDW} \to \infty$ accordingly $M_{bp} \to \infty$ and CDW turns out to be practically immobile.

An expression for $E_{bp}(\mathbf{P})$ is complicated in the general case and even in the case $\mathbf{P} = 0$, there are only variational estimates for it (Lakhno, 2015b). It can be assumed that the general form of the dependence of E_{bp} on $\mathbf{P}$ will be δ-shaped, in which the δ-shaped minimum will correspond to $\mathbf{P} = \mathbf{P}_{CDW}$ (Fig. 7.1).

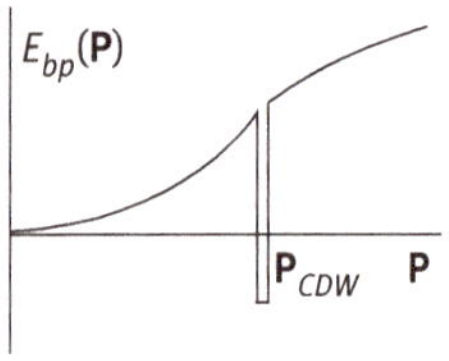

Fig. 7.1: The suggested TI bipolaron spectrum for a charge density wave. It looks as the roton spectrum, but is sharper at $\mathbf{P}_{CDW}$ due to the Kohn anomaly.

Having such a spectrum, TI bipolarons will pass on to the state with a minimum energy $E_{bp}(\mathbf{P}_{CDW})$, forming a single CDW of paired states which is determined by expression (7.1.2).

It follows that for T_{CDW}, $T_{bp} > T_c$, where $T_{CDW} = |E_{bp}(\mathbf{P}_{CDW}) - E_{bp}(0)|$, $T_{bp} = |E_{bp}(0) - 2E_p(0)|$, $E_p(0)$ is the polaron ground-state energy, the pseudogap precedes a SC. If in this case a condition $T_{CDW} > T_{bp}$ is fulfilled, the pseudogap represents a coherent pseudophase, while for $T_{bp} > T_{CDW}$, a certain non-coherent phase of free pairs precedes the coherent pseudophase. If the inequality T_{CDW}, $T_{bp} > T_c$ is not fulfilled, then the smallest quantity (T_{CDW} or T_{bp}) becomes equal to T_c. In any case a superconducting phase coexists with the pseudogap one which disappears at $T = 0$, when all the pairs are in a Bose condensate and the amplitude of the CDW vanishes.

Formation of CDW leads to violation of translational invariance up to a temperature equal to the temperature of the superconducting transition T_c. At $T < T_c$ the inhomogeneous state with CDW ceases to be thermodynamically advantageous and a coexisting translation-invariant phase is formed with restored symmetry, while TI bipolarons pass on to the Bose-condensate state. The energy advantage of such a phase follows from expressions (7.1.5) and (7.1.6), according to which a homogeneous Bose condensate has a lower energy under the condition:

$$\mathbf{P} < M_{bp}\sqrt{2\omega_0(\mathbf{P},k)/M}. \tag{7.1.7}$$

It should be noted that the scenario considered is in many respects close to the Fröhlich model of superconductivity (Fröhlich, 1954). In the Fröhlich model, it was assumed that two electrons with opposite momenta (as in the BCS) on the Fermi surface are coupled by an interaction with a phonon with a wave vector $\mathbf{P}_{CDW}$ ($2k_F$ in 1D case), thereby forming a charged phonon. Being bosons, such phonons in the macroscopic number can be in the same state with the wave vector $\mathbf{P}_{CDW}$ forming a CDW (Fröhlich charge wave). Such a wave, however, will not be superconducting, since the pinning which is always present in real crystals or its scattering by normal carriers will slow down such a wave. The main difference between the TI bipolaron description and the Fröhlich approach is the formation of a Bose condensate of TI bipolarons (which form a wave of Fröhlich charged phonons), which is responsible for superconductivity.

7.1.3 Comparison with the experiment

To be specific, let us consider the case of such a HTSC as YBCO. The CDW vector in YBCO lies in the *ab* plane and has two equally possible directions: along the a-axis ($\mathbf{P}_{CDW,x}$) and along the b-axis ($\mathbf{P}_{CDW,y}$), corresponding to the antinodal directions. For $T_c < T < T^*$ nonzero soft phonon modes $\omega_0(\mathbf{P}_{CDW},k)$ correspond to this direction of $\mathbf{P}_{CDW}$. The availability of CDW in these directions in YBCO was revealed in a large number of experiments, including nuclear magnetic resonance (Wu et al., 2011, 2013, 2015), resonant inelastic X-ray scattering (Ghiringhelli et al., 2012; Achkar et al., 2012; Blackburn et al., 2013), resonance scattering, and hard X-ray diffraction (Comin et al., 2014; Chang et al., 2012). The relevant softening of phonon modes during CDW formation was observed in Tacon et al. (2014).

A still greater softening of the phonon modes can be expected at $T < T_c$. The phonon mode corresponding to the nodal direction can vanish, which corresponds to the absence of a gap in the nodal direction. Experimental confirmation of this fact was obtained (Comin et al., 2014) for $Bi_2Sr_{2-x}La_xCuO_{6+\delta}$ (Bi2201) using combined methods of resonant X-ray scattering, scanning tunneling microscopy, and angle-resolved photoemission spectroscopy. In the case of YBCO with $\omega_0(\mathbf{P}_{CDW},k) = \Delta_0|\cos k_x a - \cos k_y a|$ (the absolute value of the CDW order parameter or the gap energy (Comin et al., 2015; Chowdhury and Sachdev, 2014), which in the TI bipolaron theory (Lakhno, 2018, 2019a, 2019b, 2020a) is the frequency of a renormalized phonon) according to (7.1.7), any value of $\mathbf{P}_{CDW}$ leads to instability of the CDW and the formation of the SC phase in this direction with the preservation of the pseudogap state in the antinodal direction.

It should be noted that, in the described approach, the difference between CDW and PDW theories disappears and $\mathbf{P}_{CDW} = \mathbf{P}_{PDW}$ (Hamidian et al., 2016) (the current state of affairs concerning the theory and experiment with PDW in high-temperature superconductors, ultracold atomic gases, mesoscopic devices is given in the review (Agterberg et al., 2020)).

7.2 Translation-invariant bipolarons and CDW pinning

As was mentioned in the previous section, for the first time the possibility for a CDW to move without experiencing resistance, or the possibility of "CDW superconductivity" was noted by Fröhlich (Fröhlich, 1954). It is believed that this process is hindered by CDW pinning at lattice defects and CDW scattering on normal current carriers.

Of interest is to discuss this problem on the basis of the microscopic theory of CDW considered in the previous section. According to the above results, in high-temperature superconductors, CDWs are formed in the pseudogap phase from bipolaron states due to the Kohn anomaly, forming a PDW for wave vectors $\mathbf{P} = \mathbf{P}_{CDW}$ corresponding to nesting.

Let us write $E_{bp}^{pin}(\mathbf{P})$ for the energy of a TI bipolaron in a CDW in the state of pinning, $E_{bp}(\mathbf{P})$ – for the energy of a free TI bipolaron, that is, in the state of depinning. Then, the motion of a CDW in a homogeneous electric field will correspond to Schrödinger equation which in the momentum representation will have the form:

$$-i\hbar F\frac{d\Psi(\mathbf{P})}{d\mathbf{P}} + (\Delta E(\mathbf{P}) - W)\Psi(\mathbf{P}) = 0 \tag{7.2.1}$$

where $\Psi(\mathbf{P})$ and W are the wave function and energy of a TI bipolaron in the momentum representation, $\Delta E(\mathbf{P}) = E_{bp}(\mathbf{P}) - E_{bp}^{pin}(\mathbf{P})$ is the energy of bipolaron coupling on an impurity, $F = 2eE$, where $2e$ is a bipolaron charge, E is the intensity of the applied electric field.

Solution of this equation has the form:

$$\Psi(\mathbf{P}) \sim \exp\, i\phi(\mathbf{P}), \tag{7.2.2}$$

$$\phi(\mathbf{P}) = \frac{1}{\hbar F}\left(W\mathbf{P} - \int_0^P \Delta E_{bp}(\mathbf{P}')\,d\mathbf{P}'\right).$$

Since in the case of a CDW the maximum of the phase distribution $\phi(P)$ in (7.2.2) over momenta corresponds to $\mathbf{P} = \mathbf{P}_{CDW}$, then it follows from the condition $\phi'_{\mathbf{P}=\mathbf{P}_{CDW}} = 0$ that $W = \Delta E_{bp}(\mathbf{P}_{CDW})$, and the condition of depinning takes on the form:

$$\phi(\mathbf{P} = \mathbf{P}_{CDW}) = 2\pi n, \quad n = 1, 2, 3, \ldots \tag{7.2.3}$$

Condition (7.2.3) corresponds to the fact that the wave phase remains unchanged as the momentum changes slightly. If condition (7.2.3) is not fulfilled, the phase $\phi(\mathbf{P})$ is a rapidly oscillating function **P**, which leads to oscillations of the sign of the charge density in a CDW. AS a result, as **P** changes slightly, the force acting from the field will also change the sign, thereby living the wave in the pinning.

Equation (7.2.3) yields the expression for the critical value of the field F_n, for which the depinning takes place:

$$F_n = \frac{1}{2\pi\hbar n}\left(\mathbf{P}_{CDW}\Delta E_{bp}(\mathbf{P}_{CDW}) - \int_0^{P_{CDW}} \Delta E_{bp}(\mathbf{P}')d\mathbf{P}'\right). \qquad (7.2.4)$$

It follows from (7.2.4) that F_1 is the maximum possible field at which pinning is preserved. For $F > F_1$ a stationary value of the phase is impossible and current oscillations are generated due to oscillations of the CDW. It should be noted that a failure of the pinning can also occur in a weaker field: $F_n < F_1, \quad n = 2, 3, \ldots$. The appearance of a larger number of critical field overtones is also possible. The nature of the occurrence of stochastic oscillations for $F > F_n$ is associated with the fact that when this condition is met, the CDW in the field moves with acceleration and its momentum increases, accordingly, condition (7.2.3) is no longer satisfied, which leads to slowing down of the wave until condition (7.2.3) is fulfilled. This explains the reason why the conductivity of systems with CDW does not exceed the conductivity of normal systems in which a CDW is absent.

It should be noted that pinning occurs only if the condition $\Delta E(\mathbf{P}) > 0$ is met. Otherwise, a free bipolaron state with energy $E_{bp}(\mathbf{P})$ is energetically more advantageous. In this case, in the pinning state, a bipolaron, being captured by a defect, loses its translational invariance, as a result of which its energy increases. For this reason, for the actual realization of pinning, a sufficiently large depth of the potential well of the defect is required. If this condition is not met, a TI bipolaron remains free and for $\mathbf{P} = \mathbf{P}_{CDW}$ does not scatter on phonons, since $\alpha_{CDW} \gg 1$. The absence of scattering of a TI bipolaron on optical phonons was first shown by Tulub (1961). The CDW formed by TI bipolarons in this case becomes "superconducting," in accordance with Fröhlich's hypothesis on the possibility of superconductivity based on CDW (Fröhlich, 1954).

7.3 Strange metals

According to the results presented in Section 7.1, the properties of the pseudogap phase are determined by the relations between T_{CDW}, T_{bp}, and T_c. As shown above, the condition $|E_{bp} - 2E_p| < T^*$ must be satisfied for the stability of the pseudogap phase. This means that, even though the bipolaron states are energetically less advantageous than polaron ones, their contribution into the thermodynamic properties at $T < T^*$ is dominant. Such dominance, however, may be absent in transport properties. In any case, for $T > T^*$ the dominant contribution to both the thermodynamic characteristics and the transport ones is made by TI polarons. When $T < |E_P|$ their energy lies deeper than the conducting layer of the Fermi surface, their concentration is much less than the concentration of electrons, accordingly, under the Fermi surface they represent a non-degenerate conducting fermion gas. As in the case of TI bipolarons, due to the Kohn anomaly, when the momentum of a TI polaron corresponds to nesting, the relevant EPI constant is large, which leads to their large effective mass. It follows that at a finite nesting value, TI polarons have a low velocity (poor metal). For $T > T^{**} \cong |E_P|$ TI polarons decay into ordinary electrons and phonons, and a poor (or strange) metal becomes an ordinary metal.

Since the concentration of TI polarons both in the pseudogap and in the "strange" phase is low, the electrical resistance ρ caused by them will be proportional to the number of phonons and for $\omega_0 < T$ increases linearly with temperature:

$$\rho \sim T/\omega_0, \; T > T_c. \tag{7.3.1}$$

This dependence distinguishes strange metals from ordinary ones, in which, in the region of sufficiently low temperatures $(T < 0.2\omega_0 \ll E_F)$ the electrical resistance is described by the expression (Sadovskii, 2021)

$$\rho(T) = \rho_0 + a_{ee}\left(\frac{T}{E_F}\right)^2 + a_{ep}\left(\frac{T}{\omega_0}\right)^5, \tag{7.3.2}$$

where ρ_0 is the residual resistance at zero temperature, the second term on the right-hand side is the contribution of electron–electron scattering, and the third one is the contribution of electron–phonon scattering.

Linear dependence (7.3.1) for $T > T_c$ is usually observed in overdoped cuprates $(T^* \approx T_c)$ in the whole range of temperatures $T > T_c$. The transition from a linear dependence of the resistance for $T > T^*$ to a nonlinear one for $T < T^*$ suggests a transition from the phase of a strange metal to a pseudogap phase, where CDW/PDW mechanism can dominate. This conclusion is confirmed by experiments in HTSC $La_{1.6-x}Nd_{0.4}Sr_xCuO_4$, in which superconductivity for $T > T_c$ was suppressed by an external magnetic field (Daou et al., 2009).

It should be noted that due to anisotropy of the SC gap, ρ will have the same anisotropy. In particular, one can expect a minimum in the scattering of TI polarons (bipolarons) in the nodal direction, since, according to Tulub (1961), there will be no scattering of TI polarons by optical phonons at a sufficiently large EPI constant. In the case of optimal doping, this conclusion is confirmed by experiments with overdoped HTSC $Tl_2Ba_2CuO_{6+\delta}$ (Taillefer, 2006; Abdel-Jawad et al., 2006).

Afterword

After the discovery of high-temperature superconductivity, it became clear that the BCS theory does not suit to explain this phenomenon. A large number of theories have emerged based on the non-phonon mechanism of interaction, which have been supported by various experiments.

It is widely believed that high-temperature superconductivity is a very complex phenomenon, which can be explained only if various mechanisms of an interaction between electrons and the crystal lattice are taken into account. However, this is in conflict with the epigraph to the monograph.

The approach presented in the monograph, like the BCS, is based on the electron–phonon mechanism. A method for solving the problem of the interaction of one and two particles in a phonon field is described in detail. The solutions obtained are used to develop a translation-invariant bipolaron theory of superconductivity with a strong electron–phonon interaction. It is shown that this approach is able to explain many experiments on HTSC and can be generalized to other mechanisms.

Taking into account of the above, the theory of superconductivity, developed on the basis of the electron–phonon interaction, actually should be considered as a scheme of actions for describing the properties of real materials (some comments and suggestions for future work and problems to be solved are added in Appendices J and K). The number of various compounds with HTSC properties is many thousands, and the number of publications on HTSC is many hundreds of thousands. Therefore, the construction of a microscopic theory of HTSC, apparently, should be understood as a certain ideological concept, the role of which can be played by the TI bipolaron mechanism considered in the book.

https://doi.org/10.1515/9783110786668-008

Acknowledgments

The author expresses his deep gratitude to A.V. Tulub for numerous and fruitful discussions; B.M. Smirnov for his recommendations and support in the process of book preparation; N.I. Kashirina for a long-time collaboration in the field of polaron physics. In the course of writing the book, I have profited from valuable and stimulating discussions with my colleges from Bogolyubov laboratory of theoretical physics: V.A. Osipov, S.I Vinitskii, V.S. Melezhik, and V.Yu. Yushankhai. I am especially grateful to N.M. Plakida for his valuable criticism. I am also indebted to Y.E. Lozovik for valuable comments. The author is very grateful to E.H. Lieb for e-mail discussion of his theorem on the lowest bound of polaron energy. I am also grateful to referees and editors: J.Fink, A. Bianconi, D.Agterberg, and K.J. Kapcia who indirectly contributed in the book. The author is very grateful to V.B. Sultanov for collaboration and E.V. Sobolev and A.N. Korshunova for their help in calculations, to O.B. Ivanova for the help with translation of the book and to O.V. Kudruashova for the help with book preparing.

Finally, I would like to thank the De Gruyter editors of physical sciences Kristin Berber-Nerlinger and J. Kischke for their collaboration.

https://doi.org/10.1515/9783110786668-009

Appendices

Appendix A Hamiltonian H_1

Hamiltonian H_1, involved in (2.3.2) of Chapter 2, has the form:

$$H_1 = \sum_k \left(V_k + f_k \hbar\omega_k\right)\left(a_k + a_k^+\right) + \sum_{k,k'} \frac{\mathbf{kk}'}{m} f_{k'}\left(a_k^+ a_k a_{k'} + a_k^+ a_{k'}^+ a_k\right) \\ + \frac{1}{2m}\sum_{k,k'} \mathbf{kk}' a_k^+ a_{k'}^+ a_k a_{k'}, \tag{A.1}$$

where $\hbar\omega_k$ is given by expression (2.3.5). Let us act on functional Λ_0 in (2.11.9) of Chapter 2 by operator H_1. Let us show that $\langle 0|\Lambda_0^+ H_1 \Lambda_0|0\rangle = 0$. Indeed, the action of H_1 on Λ_0 on the terms containing an odd number of operators in H_1 (i.e., on the first and second terms in H_1) will always contain an odd number of them, and the value of mathematical expectation of these terms will vanish.

Let us consider mathematical expectation of the last term in H_1:

$$\langle 0|\Lambda_0^+ \sum_{k,k'} \mathbf{kk}' a_k^+ a_{k'}^+ a_k a_{k'} \Lambda_0|0\rangle. \tag{A.2}$$

Function $\langle 0|\Lambda_0^+ a_k^+ a_{k'}^+ a_k a_{k'} \Lambda_0|0\rangle$ represents the norm of the vector $a_k a_{k'} \Lambda_0|0\rangle$ and will be positively defined for all the values of $\mathbf{k}$ and $\mathbf{k}'$. If in (A.2) we replace $\mathbf{k} \to -\mathbf{k}$, then the whole expression will change its sign and therefore (A.2) is also equal to zero. Hence, $\langle 0|\Lambda_0^+ H_1 \Lambda_0|0\rangle = 0$.

https://doi.org/10.1515/9783110786668-010

Appendix B
To solution of equation (2.3.10), Chapter 2

Following Tulub (2015), let us introduce Green's function $G_{kk'}(t)$:

$$G_{kk'}(t) = i\langle T[q_k(t), q_{k'}(0)]\rangle,$$

Fourier components of which satisfy the equation:

$$(\omega_k^2 - \omega^2)G_{kk'}(\omega) + 2\sum_{k''} \mathbf{kk}'' \sqrt{\omega_k \omega_{k''}} f_k f_{k''} G_{k''k'}(\omega) = \delta_{kk'}. \tag{B.1}$$

For $f_k = 0$, that is when an interaction is lacking, we get

$$\langle k|G_\pm^0(\omega)|k'\rangle = \frac{\delta_{kk'}}{\omega_k^2 - \omega^2 \mp i\varepsilon}.$$

Then solution of equation (B.1) has the form:

$$\langle k|G_\pm^0(\omega)|k'\rangle = \frac{\langle k|G_\pm^0|k'\rangle}{D(\omega^2)}, \tag{B.2}$$

where $D(\omega^2) = D(s)$ is determined by (2.3.19) of Chapter 2.

The spectrum of excitations of the system under consideration reckoned from the ground state energy is determined by the poles $\langle k|G_\pm^0(\omega)|k'\rangle$ on the real axis and, according to (B.2), has the form $\omega = \omega_k$, corresponding to the values of ω, for which $D(\omega^2) = 0$.

The matrices $M_\pm$ involved in Bogolubov transformation (2.11.1) of Chapter 2:

$$\begin{aligned} a_k &= (M_+)_{kk'}\alpha_{k'} + (M_-^*)_{kk'}\alpha_{k'}^+, \\ a_k^+ &= (M_+^*)_{kk'}\alpha_{k'}^+ + (M_-)_{kk'}\alpha_{k'}, \end{aligned} \tag{B.3}$$

where the repeating indexes mean the summation and are related to solutions of (2.3.10) as follows:

$$(M_\pm)_{kk'} = \frac{1}{2}\frac{\omega_k \pm \omega_{k'}}{(\omega_k\omega_{k'})^{1/2}}\langle k|\Omega_\pm|k'\rangle,$$

$$\langle k|\Omega_\pm|k'\rangle = \delta(\mathbf{k} - \mathbf{k}') + \frac{2(\mathbf{kk}')f_k f_{k'}(\omega_k\omega_{k'})^{1/2}}{(\omega_{k'}^2 - \omega_k^2 \mp i\varepsilon)D_\pm(\omega_k^2)},$$

where Ω is the Møller unitary matrix (Schweber, 1961). Unitarity of transformation (B.3) was proven in Tulub (1960).

https://doi.org/10.1515/9783110786668-011

Appendix C
Analytical properties of the function $D(s)$

Let us show that (2.5.4) and (2.5.5) follow from (2.5.1) and (2.5.2). To this end, we note that the analytic properties of $D(s)$, indicated in Tulub (1961), follow directly from (2.3.19), Chapter 2. Indeed, the pole $D(s)$ can only lie on the real axis, since due to positive definiteness of $\omega_k k^4 f_k^2$ in (2.3.19):

$$1+\frac{1}{3\pi^2}\int_0^\infty \frac{\omega_k k^4 f_k^2\left(\omega_k^2-s_0+i\varepsilon\right)}{\left(\omega_k^2-s_0\right)^2+\varepsilon^2}dk=0 \tag{C.1}$$

can be fulfilled only for $\varepsilon=0$. Besides, $D(s)$ is a monotonously growing function of s in view of the fact that $D'(s)>0$ for $s<1$ and as $s_0\to\infty$ $D(s)$ turns to unity. Therefore, $D(s)$ cannot have zeros for $-\infty<s_0<1$. Therefore, function $(s-1)D(s)$ can be presented in the form:

$$\frac{1}{(s-1)D(s)}=\frac{1}{2\pi i}\oint_{C+\rho}\frac{ds'}{(s'-s)(s'-1)D(s')}, \tag{C.2}$$

where the contour of integration in the Cauchy integral (C.2) is shown in Fig. 2.1. The integrand in (C.2) has a pole for $s'=1$ and a cut from $s'=1$ to $s'\to\infty$. Performing integration in (C.2) over the upper and lower edges of the cut, we obtain the integral equation (2.5.5).

https://doi.org/10.1515/9783110786668-012

Appendix D
Calculation of recoil energy ΔE

Let us make a detailed calculation of the quantity ΔE (2.5.7) of Chapter 2 with the use of the probe function (2.5.9).

For this purpose, to calculate the real part of $D\left(\omega_p^2\right)$ involved in (2.5.7) we use the Sukhotsky formula:

$$\frac{1}{\omega_k^2 - \omega_p^2 - i\varepsilon} = P\frac{1}{\omega_k^2 - \omega_p^2} + i\pi\delta\left(\omega_k^2 - \omega_p^2\right),$$

$$\operatorname{Re} D\left(\omega_p^2\right) = 1 + \frac{1}{3\pi^2}\int_0^\infty f_k^2 k^4 P\frac{\omega_k}{\left(\omega_k^2 - \omega_p^2\right)} dk.$$

It is convenient to represent Re D in the form:

$$\operatorname{Re} D = 1 + I_1 + I_2,$$

$$I_1 = \frac{1}{3\pi^2}\int_0^\infty f_k^2 k^4 \frac{dk}{\left(\omega_k + \omega_p\right)}, \qquad I_2 = P\frac{\omega_p}{3\pi^2}\int_0^\infty \frac{f_k^2 k^4 dk}{\left(\omega_k - \omega_p\right)\left(\omega_k + \omega_p\right)}.$$

Substituting f_k in the form of (2.5.9), we express I_1 in the form:

$$I_1 = \frac{8\alpha}{3\sqrt{2}\pi}\int_0^\infty e^{-k^2/a^2} dk - \frac{8\alpha(p^2+4)}{3\sqrt{2}\pi}\int_0^\infty \frac{e^{-k^2/a^2}}{k^2 + p^2 + 4} dk.$$

Assuming that $k/a = \tilde{k}$, in the strong coupling limit ($a \to \infty$), the quantity I_1 will be

$$I_1 = \frac{8\alpha a}{3\sqrt{2}\pi}\left[\frac{\sqrt{\pi}}{2} - \frac{\pi}{2}\tilde{p}e^{\tilde{p}^2}\left(1 - \frac{2}{\sqrt{\pi}}\int_0^{\tilde{p}} e^{-t^2} dt\right)\right].$$

Accordingly, I_2 is written as follows:

$$I_2 = P\frac{4\alpha\omega_p}{3\pi\sqrt{2}}\int_0^\infty \frac{e^{-k^2/a^2} k^2 dk}{\left(\omega_k - \omega_p\right)\left(\omega_k + \omega_p\right)}.$$

This integral can be presented in the form:

$$I_2 = I_{20} + I_{21},$$

https://doi.org/10.1515/9783110786668-013

where

$$I_{20} = \frac{16\alpha\omega_p}{3\pi\sqrt{2}}\left(1 - \frac{\omega_p - 1}{p^2+2}\right)\int_0^\infty \frac{e^{-k^2/a^2}}{k^2+p^2+4}dk,$$

$$I_{21} = \frac{16\alpha\omega_p(\omega_p-1)}{3\pi\sqrt{2}(p^2+2)} P\int_0^\infty \frac{e^{-k^2/a^2}}{k^2-p^2}dk.$$

For the integrals involved in I_{20} and I_{21}, we get

$$\int_0^\infty \frac{e^{-k^2/a^2}}{k^2+p^2+4}dk = \frac{1}{a}\left[1 - \frac{2}{\sqrt{\pi}}\int_0^{\tilde{p}} e^{-t^2 dt}\right]\frac{\pi}{2}\frac{e^{\tilde{p}^2}}{\tilde{p}},$$

$$P\int_0^\infty \frac{e^{-k^2/a^2}}{k^2-p^2}dk = -\frac{\sqrt{\pi}}{a}\frac{e^{-\tilde{p}^2}}{\tilde{p}}\int_0^{\tilde{p}} e^{t^2}dt.$$

As a result, I_2 takes the form:

$$I_2 = \frac{2}{3}\frac{\alpha a\tilde{p}}{\sqrt{2}}e^{\tilde{p}^2}\left[1 - \frac{2}{\sqrt{\pi}}\int_0^{\tilde{p}} e^{-t^2}dt\right] - \frac{4}{3}\frac{\alpha a\tilde{p}}{\sqrt{2}}e^{\tilde{p}^2}\int_0^{\tilde{p}} e^{t^2}dt.$$

Finally, Re D is

$$\mathrm{Re}\, D = 1 + \frac{4\alpha a}{3\sqrt{2\pi}}\left(1 - \tilde{p}e^{\tilde{p}^2}\int_{\tilde{p}}^\infty e^{-t^2}dt - \tilde{p}e^{-\tilde{p}^2}\int_{\tilde{p}}^\infty e^{t^2}dt\right).$$

This result reproduces the quantity determined by formula (2.5.10). For the imaginary part Im D, according to the Sukhotsky formula, we get

$$\mathrm{Im}\, D = \frac{1}{3\pi}\int_0^\infty f_k^2 k^4 \omega_k \delta\left(\omega_k^2 - \omega_p^2\right)dk = \frac{1}{6\pi}f_p^2 p^3.$$

As a result, $\left|D\left(\omega_k^2\right)\right|$ will be

$$|D|^2 = (\mathrm{Re}\, D)^2 + (\mathrm{Im}\, D)^2 = \frac{2}{9}\alpha^2 a^2\left[e^{-2\tilde{p}^2}\tilde{p}^2 + \frac{8}{2\pi}\left(1 - \tilde{p}\int_{\tilde{p}}^\infty e^{-t^2}dt - \tilde{p}e^{-\tilde{p}^2}\int_{\tilde{p}}^\infty e^{t^2}dt\right)^2\right].$$

The first term in formula (2.5.7) is easily calculated and equal to

$$\frac{1}{4\pi^2}\int_0^\infty \frac{k^4 f_k^2}{(1+Q)}dk = \frac{3}{16}a^2.$$

When calculating the second term in (2.5.7), let us single out the integral I_p:

$$I_p = \int_0^\infty e^{-k^2/a^2} \frac{k^2(\omega_k\omega_p + \omega_k(\omega_k+\omega_p)+1)}{(\omega_k+\omega_p)^2} dk.$$

As $a \to \infty$ it is equal to

$$I_p = a^3 \frac{\sqrt{\pi}}{4}\left(1 - \tilde{p}^3 e^{\tilde{p}^2} \int_{\tilde{p}}^\infty e^{-t^2} dt (2+4\tilde{p}^2) + 2\tilde{p}^4\right) = \frac{a^3\sqrt{\pi}}{4}(1-\Omega(\tilde{p})),$$

where

$$\Omega(\tilde{p}) = 2\tilde{p}\left\{(1+2\tilde{p}^2)\tilde{p}e^{\tilde{p}^2}\int_{\tilde{p}}^\infty e^{-t^2}dt - \tilde{p}^2\right\},$$

which corresponds to the expression for $\Omega(y)$ in (2.5.12).

As a result, for the second term in formula (2.5.7), we get

$$\frac{1}{12\pi^4}\frac{4\pi\alpha}{\sqrt{2}}\int_0^\infty I_p p^4 f_p^2 \frac{\omega_p}{(\omega_p^2-1)\left|D(\omega_p^2)\right|^2} dp.$$

As $a \to \infty$, this expression takes the form:

$$\frac{\alpha^2 a^4}{3\pi\sqrt{\pi}}\int_0^\infty (1-\Omega(\tilde{p}))\frac{e^{-\tilde{p}^2}}{\left|D(\omega_{\tilde{p}}^2)\right|^2} d\tilde{p} = \frac{3}{16}a^2 q,$$

where $q = q(0)$ is given by expression (2.5.12). Hence, finally for ΔE (2.5.7), we have

$$\Delta E = \frac{3}{16}a^2(1+q),$$

which corresponds to the first term on the right-hand side of (2.5.11).

Appendix E
Squeezed states

As was first shown in Tulub (1961), in order to obtain the polaron energy in the whole range of α variation, its calculation after two canonical transformations of the Fröhlich Hamiltonian (those of Heisenberg and Lee, Low, and Pines) should be carried out not by vacuum wave functions $|0\rangle$, but by $\Lambda_0|0\rangle$, which have the meaning of squeezed states. Since Λ_0 is a unitary operator, then the ground state energy corresponding to Hamiltonian $\tilde{\tilde{H}}$ (2.3.2) can be written as follows:

$$E = \langle 0|\Lambda_0^{+}\tilde{\tilde{H}}\Lambda_0|0\rangle = \langle 0|\Lambda_0^{-1}\tilde{\tilde{H}}\Lambda_0|0\rangle.$$

Hence, for the operators involved in $\tilde{\tilde{H}}$, according to (2.11.1) we get

$$\Lambda_0^{-1}a_k\Lambda_0 = \sum_{k'} M_{1kk'}\alpha_{k'} + \sum_{k'} M^{*}_{2kk'}\alpha^{*}_{k'},$$

that is, operator Λ_0 is an operator generating Bogolyubov transformations.

To clarify the physical meaning of the operator Λ_0, let us consider a one-dimensional single-mode case, for example, a one-dimensional oscillator the ground state of which is described by the wave function $\Psi(x)$. Squeezed is a state $\Psi(qx)$, where the coefficient q is different from unity. Let us introduce an operator G_0, that transforms the function $\Psi(x)$ to the form $\Psi(qx)$:

$$e^{i\lambda\hat{G}_0}\Psi(x) = \Psi(qx),$$

where λ is an arbitrary constant which is called a squeezing parameter. Let us assume that:

$$\hat{G}_0 = -i\frac{\partial}{\partial \ln x}.$$

Let us introduce the function $\tilde{\Psi}$, such that $\Psi(x) = \tilde{\Psi}(\ln x)$.

Then:

$$e^{i\lambda\hat{G}_0}\Psi(x) = e^{i\lambda\hat{G}_0}\tilde{\Psi}(\ln x) = \tilde{\Psi}(\ln x + \lambda) = \tilde{\Psi}(\ln xe^{\lambda}) = \Psi(xe^{\lambda}).$$

Putting here $q = e^{\lambda}$, we can see that the operator $\hat{G}_0$ actually performs the required transformation. The operator $G_0 = -i\partial/\partial\ln x = -ix\partial/\partial x = x\hat{p}$ is not Hermitian; therefore, it is not unitary. This is due to the fact that when squeezed by a factor of $q = e^{\lambda}$ on x, the norm of the square of the wave function Ψ decreases by the same factor. For this reason, instead of $\hat{G}_0$ it is convenient to use the Hermitian operator $\hat{G}$, related to $\hat{G}_0$ by adding the constant:

https://doi.org/10.1515/9783110786668-014

$$\hat{G} = -i\left(x\frac{\partial}{\partial x} + \frac{1}{2}\right) = \hat{x}\hat{p} - \frac{i}{2} = \frac{1}{2}(\hat{x}\hat{p} + \hat{p}\hat{x}) = -i\left(a^2 - a^{+2}\right)/2,$$

where a^+ and a are operators of the birth and annihilation of the oscillation quanta of the oscillator, and the exponent of the Hermitian operator

$$D_\lambda = e^{i\lambda\hat{G}} = e^{i\lambda/2\left(a^2 - a^{+2}\right)}, \quad D_\lambda\Psi(x) = e^{\lambda/2}\Psi\left(e^\lambda x\right),$$

is automatically unitary. The unitary operator D_λ is called the squeezing operator, which can also be written in the form used in Tulub (1961):

$$D = Ce^{a^+ A a^+},$$

where C and A are to be determined.

To do this, let us consider, using an oscillator as an example, how squeezed states can be realized. Let there be an oscillator which corresponds to the Hamiltonian $H_1 = \hat{p}^2/2m_1 + 1/2m_1\omega_1^2x^2$ with parameters fixed up to the moment $t \leq 0^-$. At the moment $t = 0$, they instantly change to the values m_2 and ω_2. Since the initial wave function has not changed with this change of the parameters, the relevant root-mean-square deviations Δx and Δp have not changed either:

$$\Delta x = \sqrt{\int_{-\infty}^{\infty} x^2\Psi_0^2 dx} = \sqrt{\frac{m_2\omega_2}{m_1\omega_1}}\sqrt{\frac{\hbar}{2m_2\omega_2}} = e^{-\lambda}\sqrt{\frac{\hbar}{2m_2\omega_2}},$$

$$\Delta p = \sqrt{\int_{-\infty}^{\infty} \hat{p}^2\Psi_0^2 dx} = \sqrt{\frac{m_1\omega_1}{m_2\omega_2}}\sqrt{\frac{\hbar\, m_2\omega_2}{2}} = e^\lambda\sqrt{\frac{\hbar\, m_2\omega_2}{2}},$$

$$\Psi_0(x) = \left(\frac{m_1\omega_1}{\pi\hbar}\right)^{1/4} \exp\left(\frac{m_1\omega_1}{2\hbar}x^2\right), \qquad e^\lambda = \sqrt{\frac{m_1\omega_1}{m_2\omega_2}}.$$

Here, $\lambda = 1$ corresponds to the vacuum value of the root-mean-square deviations of quantum fluctuations. For $\lambda \neq 1$, the dispersion of quantum fluctuations for one of the canonically conjugated variables will be less than the vacuum value, while for the other it will be greater, so that the Heisenberg uncertainty relation $\Delta x\Delta p \geq \hbar/2$ remains unchanged.

Since the operators involved in the Hamiltonian of the oscillator do not change when the parameters change, then:

$$\hat{x} = \sqrt{\frac{\hbar}{2m_1\omega_1}}\left(a_1 + a_1^+\right) = \sqrt{\frac{\hbar}{2m_2\omega_2}}\left(a_2 + a_2^+\right)$$

$$\hat{p} = -i\sqrt{\frac{m_1\omega_1\hbar}{2}}(a_1 - a_1^+) = -i\sqrt{\frac{m_2\omega_2\hbar}{2}}(a_2 - a_2^+)$$

This can be rewritten with the use of the definition of λ in the form:

$$a_1 + a_1^+ = e^{\lambda}(a_2 + a_2^+)$$

$$a_1 - a_1^+ = e^{-\lambda}(a_2 - a_2^+)$$

Expressing the operators a_1 and a_1^+ in terms of a_2 and a_2^+, we get:

$$a_1 = a_2 \operatorname{ch}\lambda + a_2^+ \operatorname{sh}\lambda$$

$$a_1^+ = a_2 \operatorname{sh}\lambda + a_2^+ \operatorname{ch}\lambda.$$

The latter relations represent the Bogolyubov transformation which preserves the commutation relations $[a_1, a_1^+] = [a_2, a_2^+] = 1$.

For the initial Hamiltonian H_1 the ground state is the vacuum function $|0\rangle_1$, which by definition vanishes when the operator a_1 acts on it:

$$a_1|0\rangle_1 = 0.$$

This condition expressed through operators a_2, takes the form:

$$(a_2 \operatorname{ch}\lambda + a_2^+ \operatorname{sh}\lambda)|0\rangle_1 = 0.$$

It is easy to check that the solution to this equation is the function:

$$|0\rangle_1 = N(\lambda)\exp\left(-\frac{1}{2}a_2^+ a_2^+ \operatorname{th}\lambda\right)|0\rangle_2.$$

This can be seen by passing from the operators a^+, a to Fock representation in which the operator a^+ is put into correspondence with c-number $\bar{a}$, and operator a – the derivative $d/d\bar{a}$.

This is the sought-for Tulub form of the squeezed operator. It can be shown that the normalization constant $N(\lambda)$ in this case is equal to: $N(\lambda) = 1/\sqrt{\operatorname{ch}\lambda}$.

The theory of squeezed states has found wide application in quantum optics. In the simplest single-mode regime (such regimes are realized in resonators), the energy of the electromagnetic field is described by a one-dimensional oscillator considered here, in which the role of the coordinate is played by the electric field, and the role of the momentum is played by the magnetic field. Thus, the use of squeezed states in quantum optics can suppress fluctuations of either electric or magnetic fields, which is extremely important for precision measurements, for example, of gravitational waves.

Appendix F
On the "incompleteness" of the translation-invariant polaron theory

In Klimin and Devreese (2012), it was concluded that the translation-invariant continual polaron theories by Tulub and Gross (Tulub, 1961; Porsch and Röseler, 1967) are "incomplete." This conclusion in Klimin and Devreese (2012) was made on the basis of Lakhno (2012b), which reproduces the results of Tulub's (1961) theory. Thus, in Klimin and Devreese (2012), the results of Tulub (1961) are also questioned.

In Klimin and Devreese (2012), it is argued that the total energy functional obtained by Tulub must contain an additional term $\delta E_R^{(PR)}$, which arises if an external cutoff is introduced into the considered continuous model. Such external cutoff was discussed in Lakhno (2012b). For this purpose, in the total energy functional, the integration over the phonon wave vectors was not carried out in infinite limits, but was limited to a certain finite value q_0. In Lakhno (2012b), it was pointed out that in the limit $q_0 \to \infty$ the quantity $\delta E_R^{(PR)} \to 0$. In contrast to this, in Klimin and Devreese (2012), an expression was given for $\delta E_R^{(PR)}$, which does not depend on q_0 and does not vanish as $q_0 \to \infty$.

Let us show that actually the value of $\delta E_R^{(PR)}$ depends on q_0 and for fixed other values of the parameters of the system $\delta E_R^{(PR)} \to 0$ as $q_0 \to \infty$. We will proceed from the expression for $\delta E_R^{(PR)}$ (Lakhno, 2012b), which is also the initial one in Klimin and Devreese (2012):

$$\begin{aligned}
\delta E_R^{(PR)}(q_0) &= \frac{3\hbar}{2}\left(\Omega_{q_0} - \omega_{q_0}\right),\\
\Omega_{q_0} &= \left\{\omega_{q_0}^2 + \int_0^1 d\eta \int_0^{q_0} dq \frac{\hbar q^4 f(q)\omega_q}{3\pi^2 m} \frac{2\mathrm{Re}F(\omega_q + i\delta) + \left|F(\omega_q + i\delta)\right|^2}{\left|1 + F(\omega_q + i\delta)\right|^2}\right\}^{1/2}\\
F(z) &= \eta \frac{\hbar}{6\pi^2 m}\int_0^{q_0} dq q^4 f^2(q)\left(\frac{1}{\omega_q + z} + \frac{1}{\omega_q - z}\right),\\
f_q &= -\left(V_q/\hbar\omega_0\right)\exp\left(-q^2/2a^2\right),
\end{aligned} \tag{F.1}$$

where $\omega_q = \omega_0 + \hbar q^2/2m$, ω_0 is the frequency of an optical phonon, V_q is a matrix element of EPI, and a is a variation parameter. In the limit $\delta \to 0$ from (F.1), we get

$$\Omega_{q_0}^2 - \omega_{q_0}^2 = \int_0^{q_0} dq \frac{\hbar q^4 f(q)^4 \omega_q}{3\pi^2 m}. \tag{F.2}$$

https://doi.org/10.1515/9783110786668-015

With the use of expression for V_q: $V_q = \left(2\pi\sqrt{2}\right)^{1/2} \alpha^{1/2} \sqrt[4]{\hbar^4 \omega_0^2/m} \times (1/q)$. where α is a constant of electron–phonon coupling, it follows from (E.1) and (E.2) that:

$$\Omega_{q_0}^2 - \omega_{q_0}^2 = \int_0^{q_0} dq \frac{\hbar q^4 f(q)^4 \omega_q}{3\pi^2 m}.$$

$$\Delta = \frac{2\sqrt{2}}{3\pi} \alpha \omega_0^2 \left(\frac{\hbar a_0^2}{m\omega_0}\right)^{3/2} N_1 \left(1 + \hbar a^2 N_2 / 2m\omega_0 N_1\right),$$

$$N_1 = \int_0^{q_0/a} dx x^2 \exp(-x^2), \qquad N_2 = \int_0^{q_0/a} dx x^4 \exp(-x^2). \tag{F.3}$$

Let us consider various limiting cases of expression (F.3). In the limit of small values a of the cutoff parameter: $q_0 \ll a$ – from (E.3), we obtain:

$$\Delta = \frac{2\sqrt{2}}{9\pi} \alpha \omega_0^2 \left(\frac{\hbar q_0^2}{m\omega_0}\right)^{3/2} \left[1 + \frac{3}{10}\left(\frac{\hbar q_0^2}{m\omega_0}\right)\right]. \tag{F.4}$$

From (F.3) and (F.4), it follows that in this limit $\delta E_R^{(PR)}$ is independent of the parameter a. For $q_0 << a$, expression (F.3) yields:

$$\delta E_R^{(PR)} = \frac{3\hbar}{2} \omega_{q0} \left(\sqrt{1 + q_{0c}^4 / q_0^4} - 1\right), \tag{F.5}$$

The quantity q_{0c} has the meaning of the quantity for which the integrand in the Tulub functional has a maximum (Tulub, 1961; Lakhno, 2010b). From (F.5), it follows that in the limit $q_0 << q_{0c}$:

$$\delta E_R^{(PR)} = \frac{3\hbar}{4} \omega_{q0} q_{0c}^4 / q_0^4, \tag{F.6}$$

that is, for large q_0 the quantity $\delta E_R^{(PR)} \approx q_0^{-2}$. For this reason, Klimin and Devreese (2012) decided that the theory by Porsch and Röseler (1967) goes over to the theory by Tulub (1961) with $\delta E_R^{(PR)} = 0$. It is easy to see, however, that this is not the case. Indeed, in the Tulub total energy functional obtained by Porsch and Röseler (1967), the upper limit of integration must satisfy the condition $q_0 > q_{0c}$, that automatically leads to a dependence of the polaron energy on the coupling constant of the form $E \propto \alpha^{4/3}$ (Lakhno, 2012b, 2012c). Thus, the theory of cutoff by Porsch and Röseler does not in any case go over into the theory of Tulub. The use of such an inconsistent theory led Klimin and Devreese (2012) to the conclusion that Tulub’s (1961) theory is “incomplete”.

To sum up, we will formulate the final conclusions. It is shown in the work that the conclusion by Porsch and Röeseler that the quantity $\delta E_R^{(PR)}$ vanishes as $q_0 \to \infty$ is correct. This result, however, does not lead to Tulub’s theory. Hence, the statement about the incompleteness of the Tulub total energy functional made by Klimin and Devreese is incorrect. The calculations (Lakhno, 2010b; Lakhno, 2012a; Kashirina et al., 2012), based on Tulub (1961), are beyond doubt.

Appendix G
Transition from discrete to continuum Holstein Hamiltonian

The Holstein Hamiltonian for an electron in a uniform molecular chain can be written as

$$H = -v\sum_n \left(c_n^+ c_{n+1} + c_{n+1}^+ c_n\right) + g\sum_n c_n^+ c_n\left(b_n^+ + b_n\right) + \sum_n \hbar\omega_0\left(b_n^+ b_n + \frac{1}{2}\right), \tag{G.1}$$

where v is a matrix element of the electron transition between neighboring sites, c_n^+, c_n are the operators of the birth and annihilation of an electron on the nth site, and b_n^+, b_n are the operators of the birth and annihilation of the oscillation quanta on the nth site.

In the case of weak coupling ($g/\hbar\omega_0 \ll 1$), the solutions of (G.1) are Bloch waves and the ground state for the energy in the second order of the perturbation theory takes the form (Klamt, 1988)

$$E_k = -2v - \frac{g^2/\hbar\omega_0}{\sqrt{1 + 4v/\hbar\omega_0}}. \tag{G.2}$$

For $g/\hbar\omega_0 \gg 1$, Holstein considered two limit cases: small-radius polaron ($v \ll \hbar\omega_0$) and large-radius polaron ($v \gg \hbar\omega_0$)[1]

Solutions in the case of small-radius polaron are translation-invariant states:

$$|\psi_k\rangle = \frac{1}{\sqrt{N}}\sum_n e^{ikn} c_n^+ e^{g/\omega_0\left(b_n^+ - b_n\right)}|0\rangle, \tag{G.3}$$

which correspond to the energy spectrum

$$E_k = -2ve^{-g^2/\hbar^2\omega_0^2}\cos ka_0 - g^2/\hbar\omega_0. \tag{G.4}$$

To pass on to the limit of large-radius polaron, in (G.1) instead of c_n^+, c_n we will use $|n\rangle\langle 0|$, $|0\rangle\langle n|$ and the operators of birth and annihilation of phonons with the momentum k: a_k^+, a_k instead of b_n^+, b_n.

Hence, we will use the relations:

$$\begin{aligned} &c_n^+ = |n\rangle\langle 0|, \quad c_n = |0\rangle\langle n|, \\ &b_n^+ = \frac{1}{\sqrt{N}}\sum_k a_k^+ e^{-ikna_0}, \quad b_n = \frac{1}{\sqrt{N}}\sum_k a_k^+ e^{ikna_0}. \end{aligned} \tag{G.5}$$

1 The problem of a small-radius polaron at $T = 0$ was first solved by Tyablikov (1952).

https://doi.org/10.1515/9783110786668-016

As a result (G.1) takes the form:

$$H = -v\sum_{n}(|n\rangle\langle n+1| + |n+1\rangle\langle n|) + \frac{g}{\sqrt{N}}\sum_{n,k}\left(a_k e^{ikna_0} + a_k^+ e^{-ikna_0}\right)|n\rangle\langle n| + \sum_k \hbar\omega_0 a_k^+ a_k. \tag{G.6}$$

Let us choose the wave function $|\psi\rangle$ in the form:

$$|\Psi\rangle = \sum_n \psi_n |n\rangle. \tag{G.7}$$

As a result, for the averaged Hamiltonian, we get $\bar{H} = \langle\Psi|H|\Psi\rangle$

$$\bar{H} = -v\sum_n\left(\psi_n^*\psi_{n+1} + \psi_n\psi_{n-1}^*\right) + \frac{g}{\sqrt{N}}\sum_{n,k}|\psi_n|^2\left(a_k e^{ikna_0} + a_k^+ e^{-ikna_0}\right) + \sum_k \hbar\omega_0 a_k^+ a_k. \tag{G.8}$$

In the case of a large-radius polaron:

$$\psi_{n\pm1} \approx \psi_n \pm \frac{\partial\psi_n}{\partial na_0}a_0 + \frac{1}{2}\frac{\partial^2\psi_n}{\partial(na_0)^2}a_0^2. \tag{G.9}$$

Having introduced a continuous variable $x = na_0$ and passed on in (G.8) from summation to integration we obtain:

$$\bar{H} = \int \Psi^* H \Psi dx,$$

$$H = -\frac{\hbar^2}{2m}\Delta_x + \frac{g}{\sqrt{N}}\sum_k\left(a_k e^{ikx} + a_k^+ e^{-ikx}\right) + \sum_k \hbar\omega_0 a_k^+ a_k, \tag{G.10}$$

where $m = \hbar^2/2va_0^2$, that is, Hamiltonian (3.2.1), Chapter 3.

While for the discrete case the exact solution for Hamiltonian (G.1) in the case of strong coupling is known and determined by equations (G.2) and (G.3), in the continuum limit the exact solution of the Hamiltonian (G.9) for $g/\hbar\omega_0 \gg 1$ is unknown.

The energy value found in (3.5.8) of Chapter 3 for $g/\hbar\omega_0 \gg 1$ is currently the lowest (Fig. G.1).

It is interesting to compare the asymptotic expressions (3.2.6) and (3.5.9) of Chapter 3 with some real system. For the Holstein energy (3.2.6), such a comparison was performed using computational experiments with a classical molecular chain of the DNA type (Lakhno and Korshunova, 2010). These results are in good agreement. Comparison with energy (3.5.9) can be carried out only for a quantum molecular chain, using, for example, the quantum Monte Carlo method.

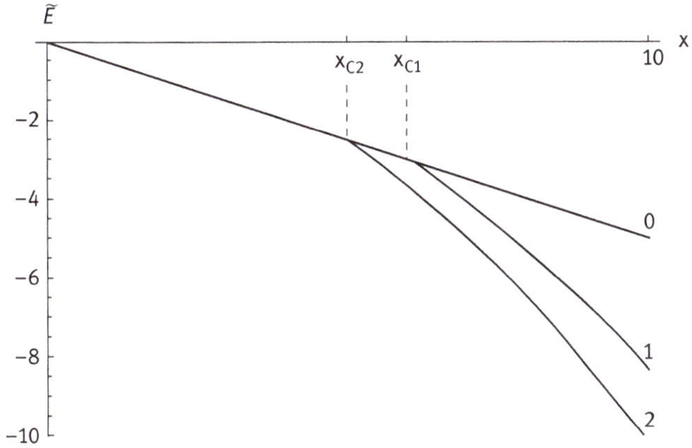

Fig. G.1: Dependence of the energy $\tilde{E} = E/\hbar\omega_0$ on the coupling constant g_c: **0)** weak coupling in the continuum limit $(v/\hbar\omega_0 >> 1)$: $\tilde{E}_{weak} = -0.5x$, $x = \sqrt{\hbar\omega_0/vg_c^2}$, $g_c = g/\hbar\omega_0$, (3.4.5), Chapter 3; **1)** strong coupling: $\tilde{E}_{strong} = -0.08333x^2$, corresponding to eq. (3.2.6), Chapter 3; **2)** strong coupling: $\tilde{E}_{strong} = -0.10185x^2$, corresponding to eq. (3.5.9), Chapter 3. The values $x_{C1} \approx 6$ and $x_{C2} \approx 5$ $g_{C1} \approx 2.45\sqrt[4]{v/\hbar\omega_0}$ and $g_{C2} \approx 2.2\sqrt[4]{v/\hbar\omega_0}$ correspond to (3.5.10), Chapter 3.

Appendix H
On the screening interaction of TI bipolarons

In Section 4.4 (Chapter 4), it was shown that, due to screening, the contribution of the interaction between bipolarons in an electron gas is small and their spectrum differs from the case of the absence of such an interaction only slightly; in this case, an approximate expression for V_k was considered. It is of interest to show that the use of an exact expression for V_k does not change the conclusions made. For this purpose, we consider the exact expression for V_k in the formulae for spectra (4.4.9) and (4.4.10):

$$V_k = \frac{4\pi e^2}{k^2 \varepsilon(k)},$$

where $\varepsilon(k)$ is a dielectric permittivity of the system considered which is determined by the general relation:

$$\varepsilon(k) = \frac{\rho_e(k)}{\rho_e(k) + \rho_i(k)}.$$

$\rho_e(k)$ is an external charge source, $\rho_i(k)$ is an internal charge induced by this source. Accordingly:

$$\frac{1}{\varepsilon(k)} = 1 + \frac{\rho_i}{\rho_e}, \qquad \rho_i = \sum_n^N \rho_{i,n},$$

where $\rho_{i,n}$ is a charge induced by the nth type of polarization. Taking into account that the inequality:

$$1 + \frac{\rho_i}{\rho_e} = 1 + \frac{\rho_{i_1}}{\rho_e} + \cdots + \frac{\rho_{i_N}}{\rho_e} < \left(1 + \frac{\rho_{i_1}}{\rho_e}\right) \cdots \left(1 + \frac{\rho_{i_N}}{\rho_e}\right)$$

(Bernoulli inequality) is fulfilled if all ρ_n have the same sign and $(\rho_{i_n}/\rho_e) > -1$ (this corresponds to the fact that the induced charge ρ_i is opposite in sign to ρ_e and in absolute value is less than the external source), the dielectric permittivity will be

$$\frac{1}{\varepsilon(k)} < \frac{1}{\varepsilon_1(k)} \cdots \frac{1}{\varepsilon_N(k)},$$

https://doi.org/10.1515/9783110786668-017

where $\varepsilon_n(k)$ are dielectric permittivities corresponding to different types of polarization. For this reason:

$$V_k = \frac{4\pi e^2}{k^2 \varepsilon(k)} < \frac{4\pi e^2}{k^2} \prod_n \frac{1}{\varepsilon_n(k)}.$$

Hence, even for estimation of $V(k)$ from overhead, the contribution of this term into the spectrum, according to the results of Section 4.4 (Chapter 4), turns out to be small.

Appendix I
Kohn anomaly

When describing the Kohn anomaly, one usually proceeds from the Fröhlich Hamiltonian of the form (Grimwall, 1981)

$$H = \sum_k \varepsilon(\mathbf{k}) c_{\vec{k}}^+ c_{\vec{k}} + \sum_q \hbar\tilde{\omega}(\mathbf{q}) b_{\vec{q}}^+ b_{\vec{q}} + \sum_{k,q} g(\mathbf{q}) c_{\vec{k}+\vec{q}}^+ c_{\vec{k}} (b_{-\vec{q}}^+ + b_{\vec{q}}), \tag{I.1}$$

where the first term corresponds to free electron gas; c_k^+ and c_k are the operators of the birth and annihilation of an electron with energy $\varepsilon(\mathbf{k})$. The second term corresponds to the lattice Hamiltonian; $b_{\vec{q}}^+$ and $b_{\vec{q}}$ are the operators of creation and annihilation of lattice vibrations with energy $\tilde{\omega}(\mathbf{q})$. The third term describes the interaction of electrons with the lattice; $g(\mathbf{q})$ is the matrix element of the interaction.

Renormalization of phonon frequencies corresponding to (I.1), according to Grimwall (1981), is determined by the expression:

$$\omega^2(\mathbf{q}) = \tilde{\omega}^2(\mathbf{q}) + 2\tilde{\omega}(\mathbf{q})|g(\mathbf{q})|^2 \mathrm{Re}[\chi(\mathbf{q})], \tag{I.2}$$

where $\omega(\mathbf{q})$ are phonons renormalized by the interaction with the electron gas whose polarizability is determined by $\chi(\mathbf{q})$.

Kohn anomaly describes vanishing of renormalized phonon modes $\omega(\mathbf{q})$ for $\mathbf{q} = \mathbf{P}_{CDW}$.

In the TI bipolaron theory of SC (Lakhno, 2018; Lakhno, 2019a; Lakhno, 2019b; Lakhno, 2020a), it is assumed that bipolarons are immersed in electron gas. The properties of such bipolarons are also described by the Fröhlich Hamiltonian of the form (I.1), but with the field of already renormalized phonons with energies $\omega(\mathbf{P}, q)$ and, accordingly, the matrix element of the interaction $V(\mathbf{q})$ instead of $g(\mathbf{q})$.

It should be noted that the spectral equation (7.1.6) of Chapter 7 is independent of the form of $V(\mathbf{q})$.

An experimental proof of the presence of renormalized phonons with zero energy in layered cuprate HTSC is the absence of a gap in the nodal direction, which, according to the definition in the TI bipolaron theory of SC, is the phonon frequency.

https://doi.org/10.1515/9783110786668-018

Appendix J
Some comments and suggestions for future work

1 BCS canonical transformation

The main difference between the approaches of TI bipolaron and BCS to HTSC is that in BCS in Hamiltonian (4.1.3) of Chapter 4, the canonical transformation of the form is used (see, e.g., Madelung (1972)):

$$\tilde{H} = e^{-S} H e^{S}, \tag{J.1}$$

$$S = \sum_{p,q} V_q \left(\alpha a^+_{-q} + \beta a_q \right) c^+_{p+q} c_p, \tag{J.2}$$

$$\alpha^{-1} = \varepsilon_p - \varepsilon_{p+q} - \hbar\omega(q), \quad \beta^{-1} = \varepsilon_p - \varepsilon_{p+q} + \hbar\omega(q), \tag{J.3}$$

which eliminates from Hamiltonian (4.1.3) the phonon variables. In the approach developed in Chapters 2–4 in contrast to BCS, the electron variables from initial Hamiltonian are removed by canonical transformations S_1, S_2, and S_3. Thus, the Hamiltonian (4.2.10) obtained in Chapter 4 contains only phonon variables. The study of such "phonon" Hamiltonian is more simple then "electron" BCS Hamiltonian.

2 Magnetic polaron

It was claimed many times that the results obtained for the phonon field can be easily transferred to magnon one. To demonstrate this, we can consider the s–d(f) electron magnon interaction in single axis antiferromagnetic (AF) which was considered in Lakhno (2011) to describe magnetic polarons:

$$\hat{H} = \frac{1}{2m}\left(\hat{P} + \frac{e}{c}\mathbf{A}\right)^2 + \frac{Q}{\sqrt{N}} \sum \left(a_k e^{ikr} + c.c.\right) + \sum \hbar\omega_M a_k^+ a_k,$$

$$Q = \frac{A}{4} \, \frac{(SH_{EA})^{1/2}\left(1 - H^2/H_E^2\right)^{3/4}}{4\left[H_E\left(1 - H^2/H_E^2\right) + 2H_A\right]^{1/2}},$$

$$\hbar\omega_M = \mu H_{EA}\left(1 - H^2/H_E^2\right),$$

$$H_{EA} = \sqrt{H_E H_a},$$

which has the same form as polaron Hamiltonian (1.2.1) in Introduction, where H_E is the exchange field of AF sublattices collapse, H_A is the field of magnetic anisotropy of single axis AF, $\mu = \mu_B = (e\hbar)/(2m_0 c)$ is Bohr magneton, S is the value of magnetic atom spin, A is the *s-d(f)* exchange constant.

https://doi.org/10.1515/9783110786668-019

For the case of magnetic bipolaron, there is the problem that for singlet state the total input of electron–magnon interaction will be zero since the total bipolaron spin in singlet state will be zero. So, in this case, the triplet magnetic bipolaron needs to be investigated.

As the most of HTSC have magnetic order the more general TI bipolaron theory where for example triplet bipolaron will be stabilized by both EPI and electron-magnon (paramagnon in pseudogap phase) interactions is required.

It is important that such triplet magnetic bipolarons will undergo Bose condensation when their spin is directed along the external magnetic field, which leads to the appearance of additional magnetization.

3 Thermodynamics of strange metals

According to Section 7.3 (Chapter 7) at $T > T^*$, where T^* is the temperature of the transition to the pseudogap phase, the main current carriers are TI polarons formed after the decay of TI bipolarons. Scattering of such a nondegenerate gas, which is a phase of a strange metal, by phonons leads to a linear temperature dependence of the resistance. Thus, the destruction of the strange metal phase must occur at temperature $T^{**} \sim |E_p|$, at the destruction temperature of TI polarons. The thermodynamics of the strange metal phase can be constructed if we take into account that the decay of TI polarons occurs with the formation of a free electron and a phonon. Statistical sums for TI polarons Z_p, free electrons Z_{el}, and phonons Z_{ph} are:

$$Z_p = \left[e^{-(\omega_0 + E_p)/T} \left(\frac{2\pi m T}{h^2} \right)^{3/2} \frac{eV}{N_p} \right]^{N_p}, \tag{J.3.1}$$

$$Z_{el} = \left[\left(\frac{2\pi m T}{h^2} \right)^{3/2} \frac{eV}{N_p} \right]^{N_p}, \tag{J.3.2}$$

$$Z_{ph} = \left[\frac{e^{-\omega_0/2T}}{1 - e^{-\omega_0/T}} \right]^{N_p}. \tag{J.3.3}$$

Accordingly, the condition for the decay of TI polarons will take the form:

$$Z_{el} Z_{ph} \geq Z_p. \tag{J.3.4}$$

Relations (J.3.1)–(J.3.4) lead to the condition:

$$T \ln \frac{T}{\omega_0} \geq |E_p|, \tag{J.3.5}$$

which determines the Lambert function. That is the study of thermodynamic properties of the strange metal phase can be carried out similarly to the case of the pseudogap phase (Section 4.9).

The described scenario of the transition of the strange metal phase to the normal phase leads to the idea of rather a blurred transition, whose properties are poorly understood as yet. Notice that when

$$|E_p| > |E_{bp} - 2E_p| \tag{J.3.6}$$

the transition temperature T^{**} determined from the condition of equality (J.3.5) from the strange metal phase to the normal metal phase will be higher than the temperature of the pseudogap phase: $T^{**} > T^{*}$. In the case of the opposite sign in inequality (J.3.6) the strange metal phase will be absent, since TI bipolarons in this case will directly fall apart into free electrons and phonons.

4 Exciton in a magnetic field

After passing on to variables (2.15.2) in Chapter 2 for the exciton Hamiltonian in a magnetic field instead of (2.15.3) we get:

$$\hat{H} = \frac{1}{2\mu}\left(\hat{P}_r - \frac{m_2 - m_1}{M}\frac{e}{c}\mathbf{A}_r - \frac{e}{c}\mathbf{A}_R\right)^2 + \frac{1}{2M}\left(\hat{P}_R - \frac{e}{c}\mathbf{A}_r\right)^2 +$$
$$\sum_k V_k a_k [\exp(i\mathbf{k}(\mathbf{R} + m_2\mathbf{r}/M)) - \exp(i\mathbf{k}(\mathbf{R} - m_1\mathbf{r}/M))] + H.c. + \sum \hbar\omega_k^0 a_k^+ a_k - \frac{e^2}{\varepsilon_\infty r} \tag{J.4.1}$$

Application of the Heisenberg transformation to this Hamiltonian gives

$$\hat{\hat{H}} = \frac{1}{2\mu}\left(\hat{P}_r - \frac{m_2 - m_1}{M}\frac{e}{c}\mathbf{A}_r - \frac{e}{c}\mathbf{A}_R\right)^2 + \frac{1}{2M}\left(\sum_k \mathbf{k} a_k^+ a_k - \frac{e}{c}\mathbf{A}_r\right)^2 +$$
$$\sum_k V_k a_k [\exp(i\mathbf{kr}m_2/M) - \exp(i\mathbf{kr}m_1/M)] + H.c. + \sum \hbar\omega_k^0 a_k^+ a_k - \frac{e^2}{\varepsilon_\infty r} \tag{J.4.2}$$

We will seek the solution of the stationary Schrödinger equation corresponding to the Hamiltonian (J.4.2) in the form:

$$\Psi_H(r, R, \{a_k\}) = \phi(R)\psi(r)\theta(R, \{a_k\}), \tag{J.4.3}$$

where $\theta(R, \{a_k\})$ is determined by expression (2.11.13) in Chapter 2. Averaging of (J.4.3) over $\phi(R)\psi(r)$ yields:

$$\bar{\bar{H}} = \frac{1}{2M}\left(\sum_k \hbar \mathbf{k} a_k^+ a_k\right)^2 + \sum_k \hbar\tilde{\omega}_k a_k^+ a_k + \sum_k \bar{V}_k\left(a_k + a_k^+\right) + \bar{T} + \bar{\Pi} + \bar{U} \qquad (\text{J}.4.4)$$

where

$$\hbar\tilde{\omega}_k = \hbar\omega_k^0 + \frac{\hbar e}{Mc}\langle\Psi|\mathbf{A}_r|\Psi\rangle \qquad (\text{J}.4.5)$$

$$\bar{T} = \frac{1}{2\mu}\langle\Phi\Psi|\left(P_r - \frac{m_2 - m_1}{M}\frac{e}{c}A_r - \frac{e}{c}A_R\right)^2|\Phi\Psi\rangle,$$

$$\bar{\Pi} = \frac{e^2}{2Mc^2}\langle\Psi|A_r^2|\Psi\rangle,$$

$$\bar{U} = \langle\Psi|\frac{e^2}{\varepsilon_\infty r}|\Psi\rangle, \quad \bar{\Psi}_k = V_k\langle\Psi|\exp\left(i\mathbf{kr}m_2/M\right) - \exp(-i\mathbf{kr}m_1/M)|\Psi\rangle.$$

Hamiltonian (J.4.5) has the same structure as the exciton Hamiltonian without a magnetic field or a TI bipolaron Hamiltonian. Using this analogy, we can conclude that, as in the case of a TI bipolaron in a magnetic field, a Bose condensate of TI excitons can be formed whose temperature of the transition into a condensate is determined by the same relation as that in (2.15.27), if we replace ω_0 by $\tilde{\omega}_0 = \omega_0\left(1 - H^2/H_{\max}^2\right)$, where $H_{\max}$ is the maximum field for which the existence of a Bose condensate is possible.

The case of singlet excitons was considered above. In the case of triplet excitons, into which singlet excitons initially excited by light can pass and which, due to their lower energy, can lead to their high population, the term $2\mu_B H$ should be added to Hamiltonian (J.4.5), where μ_B is the Bohr magneton. As is known, the condensation of Bose particles with spin (triplet TI excitons) should begin with spins directed to the magnetic field, leading to sample magnetization.

5 TI bipolaron scattering on optical phonon

Like the theory of polaron scattering by optical oscillations, the theory of TI bipolaron scattering should be constructed as a theory of resonant scattering (Schultz, 1959; Low and Pines, 1955; Ziman, 1960). The Hamiltonian (2.3.2)–(2.3.5) in Chapter 2 has the structure of a bipolaron Hamiltonian if V_k is replaced by $\bar{V}_k$ in this Hamiltonian and the constants $\bar{T}$ and $\bar{U}$ are added. For this reason, to calculate the probability of bipolaron scattering by optical phonons, we can use the calculation results obtained by Tulub (1961). This calculation is based on the calculation of the matrix element of the scattering matrix S:

$$\langle\mathbf{P}, \mathbf{k}|S - 1|\mathbf{P}_0, \mathbf{k}_0\rangle = -2\pi i\delta(E - E_0)R, \qquad (\text{J}.5.1)$$

$$R = \bar{V}_{k_0}\bar{V}_k \int d^3r \Big\{ \langle\Psi|\exp(-i\mathbf{kR})\left(\hat{H} - E - \omega - i\varepsilon\right)^{-1}\exp(i\mathbf{k}_0\mathbf{R})\Psi_0\rangle$$

$$+ \langle\Psi|\exp(i\mathbf{kR})\left(\hat{H} - E_0 + \omega - i\varepsilon\right)^{-1}\exp(-i\mathbf{k}_0\mathbf{R})|\Psi_0\rangle\Big\} = = (2\pi)^3\delta(\mathbf{P} + \mathbf{k} - \mathbf{P}_0 - \mathbf{k}_0)\, M.$$

The initial momenta of the bipolaron and phonon in (J.5.1) are denoted by $\mathbf{P}_0$, $\mathbf{k}_0$, the final ones by $\mathbf{P}$, $\mathbf{k}$, $E_P = P^2/2m_{bp}$ is the kinetic energy of the bipolaron, m_{bp} is the effective mass of the bipolaron, $|\Psi_0\rangle$, $|\Psi\rangle$ are the wave functions (2.11.13) in Chapter 2 of the initial and final states of the bipolaron,

$$M = |\bar{V}_{k_0}|^2 \cdot \left(\sum_k f_k^2 \omega_k^2/\delta E_k\right)^{-1}, \tag{J.5.2}$$

$$\delta E_k = (\mathbf{k}_0 - \mathbf{k})^2/2m + \mu^{-1}\mathbf{P}_0(\mathbf{k}_0 - \mathbf{k}) - i\varepsilon,$$

$$\mu = m(1-\xi)^{-1}, \quad \xi\mathbf{P} = \sum_k \mathbf{k} f_k^2, \quad \xi = \frac{\mathbf{P}}{P^2}\sum_k \mathbf{k} f_k^2.$$

In (J.5.1), only the main term of the expansion with respect to α is taken into account. The next terms of the expansion in terms of the electron–phonon coupling constant α are of the order of $O(\alpha^{-2})$. From (J.5.1), it follows that in the limit $P_0 \to 0$, $M^2(k_0) \sim \delta(k_r - k_0)$, where k_r is the value of the resonant momentum. Substituting into (J.5.2) the probe functions Ψ and f_k, determined by (2.9.1) and (2.9.2) in Chapter 2, we express the resonant momentum k_r as follows:

$$\frac{k_r}{\sqrt{2m\omega/\hbar}} = 16\alpha^2 \left(\frac{1}{3\sqrt{\pi}}\,\frac{\sqrt{x^2 + 16y}}{x^2 + 8y}\right)^{1/2}, \tag{J.5.3}$$

where x, y are the quantities whose dependences on $\eta = \varepsilon_\infty/\varepsilon_0$ are shown in Fig. 2.4 (Chapter 2).

It follows from (J.5.3) that the right-hand side of the equality grows with α as α^2 and the phonon momentum in the lattice has the maximum value π/a_0, where a_0 is the lattice constant. For this reason, there is always some maximum value of the coupling constant α_{max}, for which equality (J.5.3) is no longer satisfied. Thus, for $\alpha > \alpha_{max}$, the TI bipolaron cannot experience scattering, becoming superconducting with respect to scattering by optical phonons.

For $\eta = 0$ from (J.5.3) with the use of Fig. 2.3 $(x(0) \approx 5.86;\ y(0) \approx 2.59)$ we approximately get:

$$\frac{k_r}{\sqrt{2m\omega/\hbar}} \approx 2.8\alpha^2. \tag{J.5.4}$$

Accordingly, for the polaron in Tulub (1961), it was obtained:

$$\frac{k_r}{\sqrt{2m\omega/\hbar}} \approx 0.04\alpha^2. \qquad (J.5.5)$$

In Tulub (1961), for a number of parameter values, an approximate estimate was given for the maximum value of the estimate of the coupling $\alpha_{\max,pol}$, for which the polaron becomes superconducting. Comparison of (J.5.4) with (J.5.5) shows that the corresponding value of $\alpha_{\max,bip}$ for a bipolaron is related to the value of $\alpha_{\max,pol}$ as $\alpha_{\max,bip}{\sim}0.12\cdot\alpha_{\max,pol}$. For particular ionic crystals considered in Tulub (1961), the estimate $\alpha_{\max,pol}{\sim}10$ was obtained. Thus, for these crystals $\alpha_{\max,bip}\approx 1$. Since a small value was obtained for $\alpha_{\max,bip}$, the resulting estimate claims only to an order of magnitude, and the calculation of the next expansion terms is an urgent problem. Nevertheless, based on the fact that for a bipolaron $\alpha_{\max,bip}\approx O(1)$ one can expect that the real values $\alpha_{\max,bip}$ lie in the region of an intermediate coupling strength.

6 Phonon and nonphonon mechanisms

HTSC compounds have a complex crystal structure with a unit cell containing a large number of atoms. As an example, let us consider layered cuprates, the superconductivity of which is due to the presence of CuO_2 planes in them (Fig. J.1), the Cu atoms in the plane have a structure of filled levels in the form: $[Ar]3d^{10}4s^1$, and oxygen O: $1s^2 2s^2 2p^4$.

In the absence of doping, oxygen takes two electrons from copper, so that the CuO plane forms an ionic lattice composed of Cr^{2+} and O^{2-} ions. In this case, copper is in the $[Ar]3d^9$state, that is, with an unfilled d-shell and an empty 4s shell, while oxygen with a completely filled p-shell is in the $1s^2 2s^2 2p^6$ state. The absence of one electron in the d-level of copper is equivalent to the presence of one hole in the d-shell of copper.

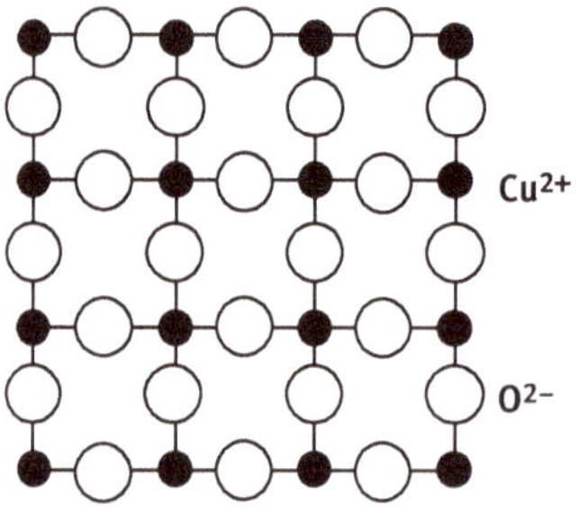

Fig. J.1: CuO_2 plane.

The presence of a hole at the d-level of copper leads to a hole in d-conduction band of copper, and since there is one hole on each copper atom, such a band will be half filled with holes. The calculation of the band structure in this case gives a wide conductivity band, as in an ordinary metal. However, this contradicts the experiment, which shows that actually in this case a dielectric is realized, and not a conductor. The reason is that the electron correlations are not taken into account in such a calculation of the band structure. In the case of a strong electron–electron interaction of copper d-electrons, the hole band changes from conducting to the state of a Mott dielectric with AF ordering of hole spins in the CuO_2 plane.

The reason for the appearance of the Mott dielectric can be interpreted as follows. In the ground state, each copper ion is in the d^9 configuration. Now suppose that one of the copper electrons jumped to another atom. Then the copper atom left by the electron will be in the d^8 configuration, while the atom on which the electron jumped will go into the d^{10} configuration. But such a jump in the case of a strong Coulomb repulsion of electrons located on the same atom will lead to a loss in the total energy of the system (which in the Hubbard model, equals to U, Chapter 3), if the gain in the kinetic energy during the jump is less than the loss in the energy of the Coulomb repulsion of two electrons on the atom where the electron jumped. Thus, in this case, it is advantageous for the electrons to remain localized on their copper atoms. In this case, the spins of neighboring copper ions will be antiparallel (i.e., they will form a Neel lattice), since their interaction is due to virtual jumps of electrons to neighboring copper ions, the spins of which, due to the Pauli principle, must be antiparallel to the spin of the electron which makes the jump. The energy of this order is called the superexchange energy J, which, by virtue of the above, will be proportional to η^2/U, where η is the jump energy (the matrix element of the transition between the sites in Hamiltonian (3.10.1)).

In real cuprates, the situation is somewhat more complicated than that described above due to the hybridization of copper d-orbitals and oxygen p-orbitals. However, the overall picture will not change if we consider the copper ion as an effective site surrounded by oxygen neighbors.

Such, for example, is the picture in the $YBa_2Cu_3O_6$ crystal. If we now dope this crystal with additional oxygen atoms that violate the stoichiometry of the crystal, as a result of which its composition will have the form $YBa_2Cu_3O_{6+x}$, then additional empty places will appear in the crystal in the d-shell of copper atoms, which corresponds to the appearance of free holes in the d-band of the Mott dielectric. If, in this case, the Coulomb repulsion of holes on the copper ion is large, then the holes that appear during doping will occupy oxygen orbitals, which, thereby, become conductive. An increase in the concentration of free holes as doping increases leads to the appearance of a hole Fermi liquid and a Fermi surface in the spectrum of free holes.

Basically, all theoretical models used to explain HTSC can be divided into two groups: phonon models and non-phonon models. Phonon models primarily include such classical continuum models as BSC and its generalization to the case of strong EPI – the Eliashberg theory. Many electron-phonon models proceed from a discrete description of the lattice in its quasi-two-dimensional approximation and the use of the strong coupling approximation. This is the Hubbard model (Section 3.10) and its many modifications. The SC theory based on small-radius bipolarons belongs to the same type of models. Superconductivity in this case is achieved due to the formation of a Bose condensate by such paired states.

The nonphonon models mainly include the models based on the presence of magnetic ordering in HTSC compounds. As a rule, superconductivity is accompanied by an AF order, although the coexistence of a superconducting phase with a ferromagnetic order is also possible.

At the end of the last century, the most popular theory with a non-phonon mechanism was the Anderson theory of resonant valence bonds (RVB) (Anderson, 1997). It is believed that the ground state of 2d AF differs from the Neel one. In it, an unpaired spin corresponds to an elementary excitation, and its transition through neighboring sites corresponds to a valence transfer. Such a transfer carries the spin, but not the charge, over the lattice sites. This moving Fermi excitation is called a spinon. The spinon, having a free spin, attracts the resulting hole and forms a bound state with it – a holon, which is a spinless charged boson. Such bosons are capable of forming a Bose condensate, but really biholons are more stable and form a Bose condensate.

At present, this approach is rather of historical interest, since subsequent calculations showed that the RVB state is energetically more disadvantageous than the Neel state and does not correspond to the experiment which confirms the Neel magnetic state in cuprates.

Since the SC theory should ultimately be based on the effect of electron (hole) pairing, models associated with the SC fluctuation mechanism are also considered as an alternative to the phonon mechanism. These ideas are often based on the idea of magnetic polarons in antiferromagnets (Nagaev, 1979; Val'kov et al., 2021; Vidmar et al., 2009). The first successful application of the theory of magnetic bipolarons to HTSC was realized by Schrieffer et al. (1988, 1989), that is, spin-bag model.

For the Bose condensation temperature of such magnetic bipolarons, Schrieffer obtained $T_C \sim 100$ K. The disadvantage of magnetic bipolaron-type models is instability of such bipolarons in the presence of even a weak external magnetic field (see (J.2)). A common problem in using the idea of magnetic polarons in SC is the low coupling energy of their bound bipolaron state.

The nonmagnetic mechanisms of HTSC can also include the exciton mechanism. An exciton is a Bose formation – a bound state of an electron and a hole (Section 2.15). Such a boson quasiparticle can act as a virtual one for the efficient interaction of current carriers in HTSC by analogy with a phonon. If the energy of the exciton is much higher than that of the phonon, then it is expected that such a mechanism can be used to obtain a sufficiently high SC temperature (Gaididei and Loktev, 1988; Weber, 1988; Varma et al., 1987).

A large number of models refer to the HTSC plasma mechanism. Plasmons in these models play the role of optical phonons. The corresponding theory of the electron-plasmon mechanism of HTSC can be constructed similarly to BCS or bipolaron superconductivity (Pashitskii and Chernousenko, 1971; Pashitskii, 1993; Takada, 1978; Davydov et al., 2020).

We have briefly considered the most popular models of SC. There exist many other mechanisms. Nevertheless, all of them use the conception of pairing with the subsequent formation of Bose condensate below T_c irrespective of the nature of the resulting attraction.

Appendix K
The problems to be solved

1. As is known, the theory of ideal Bose gas gives the continuous dependence of capacity on temperature at $T = T_{c.}$

   In contrast to this, the theory of TI bipolaron gas leads to discontinuous behavior at $T = T_c$ in accordance with experiment. The statistical properties of low-density TI bipolaron gas in Section 4.5 (Chapter 4) were calculated with the assumption that the input of d-wave part of phonon spectrum is less then s-wave one. This approach will be incorrect near the nodal direction.

   In the more exact theory, it is possible to calculate the temperature dependence of capacity taking into account the d-wave part of phonon spectrum in order to get a more detailed temperature dependence $C_V(T)$ (e.g., T^n dependence instead of $\exp(-\omega_0/T)$ dependence at low T).

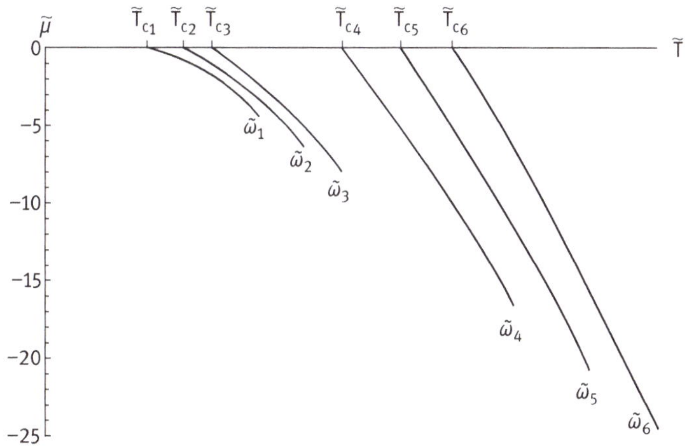

Fig. K.1: The temperature dependence of chemical potential of 3D TI bipolaron gas determined by (4.5.9), Chapter 4.

   The discontinuous behavior of $C_V(T)$ at $T = T_C$ is due to nonanalyticity of chemical potential $\mu(T)$ (formula (4.5.9), Chapter 4) which has at $T = T_c$ the nonzero derivative with respect to T (see Fig. K.1).

2. The TI bipolaron theory of pseudogap phase (Section 4.9) was developed only for the case of classical statistics. In order to describe the coexistence of SC and pseudogap phase below T_c the more general theory based on quantum statistics need to be developed. For this aim, the quantum expression for statistical sum

https://doi.org/10.1515/9783110786668-020

of TI bipolaron (formula (4.9.4), Chapter 4) and polaron (formula (4.9.5), Chapter 4) need to be calculated.

3. For a long time (and up to date), the idea that the fundamental mechanism of SC is based on magnetic interaction was grounded on neutron-scattering experiments. For YBCO, the observed neutron peak near $\omega_2 = 41\text{meV}$ (Fig. 5.13) is usually attributed to the value of SC gap (Rossat-Mignod et al., 1991; Mook et al., 1993; Dai et al., 2001). For $k{\sim}T_c^{1/2}$ this peak is linear with T_c in agreement with the most families of cuprates where the peak was found. So, with the same arguments it can be attributed to its phonon nature. The further investigation of this problem both theoretical and experimental is necessary.
4. For layered superconductors (cuprates), the 2D theory can be constructed analogously to 1D and 3D theories considered in monograph. Notice that according to TI bipolaron theory, Bose condensation is possible not only in 3D but also in 2D and 1D (Chapter 3).
5. The calculation of TI polaron (bipolaron) effective mass is an open question. The problem is how to choose the appropriate variational coefficients f_k and their dependence on total momentum P in order to get the correct result.
6. To develop the theory of TI bipolaron nonhomogeneous bose condensate is actual. For this reason, the Lee–Low–Pines and Bogolyubov transformation dependent on coordinates need to be constructed. Such theory is necessary for the description of vortex superconducting matter.
7. TI bipolaron theory with interaction determined by Hamiltonian (1.2.6) can be constructed for plasma media as in degenerate semiconductors (Lakhno, 1994) (plasmon mechanism of superconductivity).
8. There are a lot of interactions in real HTSC. The generalized TI bipolaron theory which takes into account besides electron (hole) – phonon the electron (hole) – plasmon interaction (1.2.6) and electron (hole) – magnon interaction (Appendix J.2) need to be developed.
9. The use of bipolaron Hamiltonian with interaction determined by (1.2.3) provides an opportunity to construct the theory of superconductivity for heavy nucleus and neutron stars in which in analogy with TI bipolaron theory the role of bose particles which form bose condensate is realized by dinucleons.

Abbreviations

AF	antiferromagnetic
ARPES	angle-resolved photoemission electron spectroscopy
BCS	Bardeen–Cooper–Schrieffer theory
BDW	bond density wave
BEC	Bose–Einstein condensate
CDW	charge density wave
EPI	electron–phonon interaction
HTSC	high-temperature superconductivity
IBG	ideal Bose gas
LRP	large-radius polaron
PDW	pair density wave
RVB	resonating valence band theory
SC	superconductivity
SRBP	small-radius bipolaron
STM	scanning tunneling microscopy
STS	scanning tunneling spectroscopy
TI	translation invariant
YBCO	$YBa_2Cu_3O_{6+x}$

https://doi.org/10.1515/9783110786668-021

References

Abdel-Jawad M., Kennett M.P., Balicas L., Carrington A., Mackenzie A.P., Mckenzie R.H. Hussey N.L. (2006). Anisotropic scattering and anomalous normal-state transport in a high-temperature superconductor. Nature Physics, 2, 821–825, DOI: https://doi.org/10.1038/nphys449

Aßmann M., Bayer M. (2020). Semiconductor Rydberg Physics. Adv. Quantum Technol., 3, 1900134, DOI: 10.1002/quite.201900134

Achkar A.J., Sutarto R., Mao X., He F., Frano A., Blanco-Canosa S., Le Tacon M., Ghiringhelli G., Braicovich L., Minola M., Moretti Sala M., Mazzoli C., Liang R., Bonn D.A., Hardy W.N., Keimer B., Sawatzky G.A., Hawthorn D.G. (2012). Distinct charge orders in the planes and chains of ortho-III ordered $YBa_2Cu_3O_{6+\delta}$ superconductors identified by resonant elastic X-ray scattering. Phys. Rev. Lett., 109, 167001, DOI: https://doi.org/10.1103/PhysRevLett.109.167001

Adamowski J., Gerlach B., Leschke H. (1980). Strong-coupling limit of polaron energy, revisited. Phys. Lett. A, 79, 249, DOI:10.1016/0375-9601(80)90263-7

Adamowski J., Bednarek S. (1992). Stability of large bipolarons. J. Phys.: Condens. Matter, 4, 2845, DOI: https://doi.org/10.1088/0953-8984/4/11/012

Agranovich V.M. (1968). Theory of Excitons, (M.: Nauka, 384 p.) (in Russian)

Agterberg D.F., Tsunetsugu H. (2008). Dislocations and vortices in pair-density-wave superconductors. Nature Phys., 4, 639–642, DOI: https://doi.org/10.1038/nphys999

Agterberg D.F., Seamus Davis J.C., Edkins S.D., Fradkin E., Van Harlingen D.J., Kivelson S.A., Lee P.A., Radzihovskiy L., Tranquada J.M., Wang Y. (2020). The Physics of Pair-Density waves: Cuprate Superconductors and Beyond. Annual Review of Condensed Matter Physics, 11, 231–270, DOI: https://doi.org/10.1146/annurev-conmatphys-031119-050711.

Aleksandrov L., Zagrebnov V.A., Kozlov Zh.A., Parfenov V.A., Priezzhev V.B. (1975). High energy neutron scattering and the Bose condensate in He II. Sov. Phys. JETP, 41, 915.

Alexandrov A., Ranninger J. (1981). Bipolaronic superconductivity. Phys. Rev B, 24, 1164, DOI: https://doi.org/10.1103/PhysRevB.24.1164

Alexandrov A.S., Krebs A.B. (1992). Polarons in high-temperature superconductors. Sov. Phys. Usp., 35, 345–383, DOI: 10.1070/PU1992v035n05ABEH002235

Alexandrov A.S., Mott N.F. (1994). Bipolarons. Rep. Progr. Phys., 57, 1197, DOI:https://doi.org/10.1088/0034-4885/57/12/001

Alexandrov A.S., Mott N. (1996). Polarons and Bipolarons (Singapore: World Scientific)

Alexandrov A.S., Kornilovitch P.E. (1999). Mobile Small Polaron. Phys. Rev. Lett., 82, 807, DOI: https://doi.org/10.1103/PhysRevLett.82.807

Alexandrov A.S. (1999) Comment on "Experimental and Theoretical Constraints of Bipolaronic Superconductivity in High Tc Materials: An Impossibility". Phys. Rev. Lett. 82, 2620 DOI: https://doi.org/10.1103/PhysRevLett.82.2620

Alexandrov A.S., Kabanov V.V. (1999). Parameter-free expression for superconducting T_c in cuprates. Phys. Rev. B, 59, 13628, DOI: https://doi.org/10.1103/PhysRevB.59.13628

Alexandrov A.S. (2003). Theory of Superconductivity. Weak to strong coupling, (IOP publishing, Bristol, UK)

Allcock G.R. (1956). On the polaron rest energy and effective mass. Advan. Phys., 5, 412.

Altmore F., Chang A.M. (2013). One dimensional superconductivity in nanowires, (Wiley, Germany)

Anderson P.W. (1975). Model for the Electronic Structure of Amorphous Semiconductors. Phys. Rev. Lett, 34, 953, DOI: https://doi.org/10.1103/PhysRevLett.34.953

Anderson P.W. (1997). The Theory of Superconductivity in the High-T_c Cuprates, (Princeton Series in Physics, Princeton Univ. Press, Princeton, New Jersey, USA).

https://doi.org/10.1515/9783110786668-022

Anzai H., Ino A., Kamo T., Fujita T., Arita M., Namatame H., Taniguchi M., Fujimori A., Shen Z.-X., Ishikado M., Uchida S. (2010). Energy-dependent enhancement of the electron-coupling spectrum of the underdoped $Bi_2Sr_2CaCu_2O_{8+\delta}$ superconductor. Phys. Rev. Lett., 105, 227002, DOI: 10.1103//Phys.Rev.Lett.105.227002

Apalkov V., Chakraborty T. (2006). Fractional Quantum Hall States of Dirac Electrons in Graphene. Phys. Rev. B, 73, 113103, DOI: https://doi.org/10.1103/PhysRevB.73.113103

Askerzade I. (2012). Unconventional Superconductors 153, of Springer Series in Material Science, (Springer, Berlin, Germany).

Bajaj K.K. (1974). Effect of electron-phonon interaction on the binding energy of a wannier exciton in a polarizable medium. Solid State Commun., 15, 1221–1224.

Balabaev N.K., Lakhno V.D. (1980). Soliton solutions in polaron theory. Theoretical and Mathematical Physics, 45, 936–938, DOI: https://doi.org/10.1007/BF01047152

Balabaev N.K. Lakhno V.D. (1991). Spectral properties of the polaron model of a protein. SPIE, 1403, 478, DOI: https://doi.org/10.1117/12.57276

Baranowski M., Plochocka P. (2020). Excitons in Metal-Halide Perovskites. Adv. Energy Mater, 1903659. DOI: 10.1002/aehm.201903659

Bardeen J., Cooper L.N. Schrieffer J.R. (1957). Theory of superconductivity. Phys. Rev., 108, 1175, DOI: https://doi.org/10.1103/PhysRev.108.1175

Barentzen H. (1975). Effective electron-hole interaction for intermediate and strong electron-phonon coupling. Phys. Stat. Sol. (b), 71, 245, DOI: https://doi.org/10.1002/pssb.2220710125

Batlogg B., Kourouklis G., Weber W., Cava R.J., Jayaraman A., White A.E., Short K.T., Rupp L.W., Rietman E.A. (1987). Nonzero isotope effect in $La_{1.85}Sr_{0.15}CuO_4$. Phys. Rev. Lett., 59, 912, DOI: https://doi.org/10.1103/PhysRevLett.59.912

Bednorz J.G., Müller K.A. (1986). Possible high T_c superconductivity in the Ba-La-Cu-O system. Z. Physik B - Condensed Matter, 64, 189, DOI: https://doi.org/10.1007/BF01303701

Bendele M., von Rohr F., Guguchia Z., Pomjakushina E., Conder K., Bianconi A., Simon A., Bussmann-Holder A., Keller H. (2017). Evidence for strong lattice effect as revealed from huge unconventional oxygen isotop effects on the pseudogap temperature in La_2-xSr_xCuO_4. Phys. Rev. B, 95, 014514, DOI: https://doi.org/10.1103/PhysRevB.95.014514

Benneman K.H., Ketterson J.B. (2008). Superconductivity: conventional and unconventional superconductors *1–2*. (Springer, New York)

Berg E., Fradkin E., Kivelson S.A. (2009). Charge-4e superconductivity from pair-density-wave order in certain high-temperature superconductors. Nature Phys., 5, 830–833, DOI: https://doi.org/10.1038/nphys1389

Bethe H.A. (1964). Intermediate Quantum Mechanics (New York: W A Benjamin), [Translated into Russian (Moscow: Mir, 1965)]

Bilbro L.S., Aguilar R. Valdés, Logvenov G., Pelleg O., Božović I., Armitage N.P. (2011). Temporal correlations of superconductivity above the transition temperature in $La_{2-x}Sr_xCuO_4$ probed by terahertz spectroscopy. Nat. Phys. 7, 298–302, DOI: https://doi.org/10.1038/nphys1912

Bill A., Kresin V.Z., Wolf S.A. (1998a). Isotope effect for the penetration depth in superconductors. Phys. Rev. B, 57, 10814, DOI: https://doi.org/10.1103/PhysRevB.57.10814

Bill A., Kresin V.Z., Wolf S.A. (1998b). 'The isotope Effect in Superconductors' in V.Z. Kresin (ed.), Pair Correlation in Many Fermions Systems, Plenum Press, New York, USA, 25–55. DOI: https://doi.org/10.1007/978-1-4899-1555-9

Blackburn E., Chang J., Hücker M., Holmes A.T., Christensen N.B., Liang R., Bonn D.A., Hardy W.N., Rütt U., Gutowski O., Zimmermann M.V., Forgan E.M., Hayden S.M. (2013). X-ray diffraction observations of a charge-density-wave order in superconducting ortho-II $YBa_2Cu_3O_{6.54}$ single crystals in zero magnetic field. Phys. Rev. Lett., 110, 137004 DOI: https: doi.org/10.1103/PhysRevLett.110.137004

Bloch I., Dalibard J., Zwerger W. (2008). Many-body physics with ultracold gases. Rev. Mod. Phys., 80, 885, DOI: https://doi.org/10.1103/RevModPhys.80.885

Bogolyubov N.N. (1947). On the theory of superfluidity. J. Phys. USSR, 11, 23

Bogolyubov N.N. (1950). A new form of adiabatic perturbation theory in the problem of particle coupling with a quantum field. Ukr. Mat. Zh., 2 (2), 3, (in Russian)

Bogolyubov N.N. (1958). Concerning a New Method in the Theory of Superconductivity III. JETF, 34, 73–79.

Bonn D.A., Ruixing Liang, Riseman T.M., Baar D.J., Morgan D.C., Kuan Zhang, Dosanjh P., Duty T.L., MacFarlane A., Morris G.D., Brewer J.H., Hardy W.N., Kallin C., Berlinsky A.J. (1993). Microwave determination of the quasiparticle scattering time in Y $Ba_2Cu_3O_{6.95}$. Phys. Rev. B, 47, 11314, DOI: https://doi.org/10.1103/PhysRevB.47.11314

Bohr N., Rosenfeld L. (1933). Zur Frage der Messbarkeit der elektromagnetischen Feldgrössen. Mat.-fys. Medd. Dansk Vid. Selsk. 12, no. 8 [Translated in to Russian: Izbrannye Nauchnye trudy v.2 (M.: Nauka), (1971).]

Borisenko S.V., Kordyuk A.A., Legner S., Dürr C., Knupfer M., Golden M.S., Fink J., Nenkov K., Eckert D., Yang G., Abell S., Berger H., Forró L., Liang B., Maljuk A., Lin C.T., Keimer B. (2001). Estimation of matrix-element effects and determination of the Fermi surface in $Bi_2Sr_2CaCu_2O_{8+\delta}$ systems using angle-scanned photoemission spectroscopy. Phys. Rev. B, 64, 094513, DOI: https://doi.org/10.1103/PhysRevB.64.094513

Božović I., He X., Wu J., Bollinger A.T. (2016). Dependence of the critical temperature in overdoped copper oxides on superfluid density. Nature, 536, 309, DOI: https://doi.org/10.1038/nature19061

Božović I., Wu J., He X., Bollinger A.T. (2017). On the origin of high-temperature superconductivity in cuprates. Proc. SPIE, Oxide-based Materials and Devices VIII, 10105, 1010502, DOI: https://doi.org/10.1117/12.2261512

Braunstein S.L. (2005). Squeezing as an irreducible resource. Phys. Rev. A, 71, 055801, DOI: https://doi.org/10.1103/PhysRevA.71.055801

Buckel W., Kleiner R. (2004). Superconductivity: Fundamentals and Applications. Wiley-VCH, Weinheim, 2-nd Edition.

Buymistrov V.M., Pekar S.I. (1957). The Quantum States of Particles Coupled with Arbitrary Strength to a Harmonically Oscillating Continuum. II. The Case of Translational Symmetry. Sov. Phys. JETF, 6, 977; (1957). Zh.Eksp.Teor.Fiz., 33, 1271

Camargo F., Schmidt R., Whalen J.D., Ding R., Woehl G., Yoshida Jr. S., Burgdörfer J., Dunning F.B., Sadeghpour H.R., Demler E., Kilian T.C. (2018). Creation of Rydberg Polarons in Bose gas. Phys. Rev. Lett., 120, 083401, DOI: https://doi.org/10.1103/PhysRevLett.120.083401

Carbotte J.P. (1990). Properties of boson-exchange superconductors. Rev. Mod. Phys, 62, 1027, DOI: https://doi.org/10.1103/RevModPhys.62.1027

Chainani A., Yokoya T., Kiss T., Shin S., Nishio T., Uwe H. (2001). Electron-phonon coupling induced pseudogap and the superconducting transition in $Ba_{0.67}K_{0.33}BiO_3$. Phys. Rev. B, 64, 180509, DOI: https://doi.org/10.1103/PhysRevB.64.180509

Chang J., Blackburn E., Holmes A.T., Christensen N.B., Larsen J., Mesot J., Liang R., Bonn D.A., Hardy W.N., Watenphul A., Zimmermann M.V., Forgan E.M., Hayden M. (2012). Direct observation of competition between superconductivity and charge density wave order in $YBa_2Cu_3O_{6.67}$. Nat. Phys., 8, 871–876, DOI: 10.1038/NPHYS2456

Chakraverty B.K., Ranninger J., Feinberg D. (1998). Experimental and theoretical constraints of bipolaronic superconductivity in high T_c materials: an impossibility. Phys. Rev. Lett, 81, 433, DOI: https://doi.org/10.1103/PhysRevLett.81.433

Chatterjee A., Mukhopadhyay S. (2018). Polarons and Bipolarons. An Introduction. CRC Press, Taylor & Francis Group, Boca Raton, FL, USA. DOI: https://doi.org/10.1201/9781315118635

Chen H.-D., Vafek O., Yazdani A., Zhang S.-C. (2004). Pair density wave in the pseudogap state of high temperature superconductors. Phys. Rev. Lett., 93, 187002, DOI: https://doi.org/10.1103/PhysRevLett.93.187002

Chen Q., Stajic J., Tan S., Levin K. (2005). BCS-BEC Crossover: from high temperature superconductors to ultracold superfluids. Phys. Rep., 412, 1 DOI: https://doi.org/10.1016/j.physrep.2005.02.005

Chen X.-J., Liang B., Ulrich C., Lin C.-T., Struzhkin V.V., Wu Z., Hemley R.J., Mao H., Lin H.-Q. (2007). Oxygen isotope effect in $Bi_2Sr_2Ca_{n-1}Cu_nO_{2n+4+\delta}$ (n=1,2,3) single crystals. Phys. Rev. B, 76, 140502(R) DOI: https://doi.org/10.1103/PhysRevB.76.140502

Chowdhury D., Sachdev S. (2014). Density-wave instabilities of fractionalized Fermi liquids. Phys. Rev. B, 90, 245136, DOI: https://doi.org/10.1103/PhysRevB.90.245136

Comin R., Frano A., Yee M.M., Yoshida Y., Esaki H., Schierle E., Weschke E., Sutarto R., He F., Soumyanarayanan, Yang He, Tacon M.Le., Efimov I.S., Hoffman J.E., Sawatzky G.A., Keimer B., Damascelli A. (2014). Charge order driven by Fermi-arc instability in $Bi_2Sr_{2-x}La_x CuO_{6+5}$. Science, 343, 390 DOI: 10.1126/science.1242996

Comin R., Sutarto R., He F., de Silva Neto E.H., Chanviere L., Frano A., Liang R., Hardy W.N., Bonn D.A., Yoshida Y., Eisaki H., Achkar A.J., Hawthon D.G., Keiwer B., Sawatzky G. A., Damascelli A. (2015). Symmetry of charge order in cuprates. Nature materials, 14, 796–800, DOI: https://doi.org/10.1038/nmat4295

Cooper L.N. (1956). Bound electron pairs in a degenerate Fermi gas. Phys. Rev., 104, 1189, DOI: https://doi.org/10.1103/PhysRev.104.1189

Cooper L.N., Feldman D. (eds.) (2011). BCS: 50 years (World Sci. Publ. Co, Singapore)

Cowley R.A., Woods A.D.B. (1968). Neutron scattering from liquid helium at high energies. Phys. Rev. Lett., 21, 787, DOI: https://doi.org/10.1103/PhysRevLett.21.787

Cucchietti F.M., Timmermans E. (2006). Strong-coupling polarons in dilute gas Bose-Einstein condensates. Phys. Rev. Lett., 96, 210401, DOI: https://doi.org/10.1103/PhysRevLett.96.210401

Curty P., Beck H. (2003). Thermodynamics and phase diagram of high temperature superconductors. Phys. Rev. Lett., 91, 257002, DOI: https://doi.org/10.1103/PhysRevLett.91.257002

Dai P., Mook H., Hunt R., Dogan F. (2001). Evolution of the resonance and incommensurate spin fluctuations in superconducting $YBa_2Cu_3O_{6+x}$. Phys.Rev.B, 63(5),054525, Doi: https://doi.org/10.1103/PhysRevB.63.054525

Damascelli A., Hussain Z., Shen Z.-X. (2003). Angle-resolved photoemission studies of the cuprate superconductors. Rev. Mod. Phys., 75, 473, DOI: https://doi.org/10.1103/RevModPhys.75.473

Daou R., Doiro-Leyrand N., LeBoeuf D., Li S.Y., Laliberte F., Cyr-Choiniere D., Jo Y.J., Balicas L., Yan J.-Q., Zhou J.-S., Goodenough J.B., Taillefer L. (2009). Linear temperature dependence of resistivity and change in t Fermi surface at the pseudogap critical point of a high-T superconductor. Nature Physics, 5, 31–34, DOI: https://doi.org/10.1038/NPHYS1109

Davydov A.S. (1971). Theory of molecular excitons (Plenum Press, New-York, 296 p.)

Davydov A., Sanna A., Pellegrini S., Dewhurst J.K., Shama S., Gross E.K. (2020). Ab initio theory of plasmonic superconductivity within the Eliashberg and density-functional formalisms. Phys. Rev. B, 102, 214508, Doi: https://doi.org/10.1103/PhysRevB.102.214508

Devereaux T.P., Hackl R. (2007). Inelastic light scattering from correlated electrons. Rev. Mod. Phys., 79, 175, DOI: https://doi.org/10.1103/RevModPhys.79.175

Devreese J.T. (1972). Polarons in ionic crystals and polar semiconductors. (Amsterdam: North-Holland)

Devreese J.T., Peeters F.M. (1984). Polarons and excitons in polar semiconductors and ionic crystals (N.Y.: Plenum)

Devreese J.T., Alexandrov A.S. (2009). Fröhlich polaron and bipolaron: recent developments. Rep. Prog. Phys., 72, 066501, DOI: https://doi.org/10.1088/0034-4885/72/6/066501
Drechsler M., Zwerger W. (1992). Crossover from BCS superconductivity to Bose condensation. Ann. Phys., 1, 15, DOI: https://doi.org/10.1002/andp.19925040105
Drozdov A.P., Eremets M.I., Troyan I.A., Ksenofontov V., Shylin S.I. (2015). Conventional superconductivity at 203 kelvin at high pressures in the sulfur hydrite system. Nature, 525, 73, DOI: https://doi.org/10.1038/nature14964
Dukelsky J., Khodel V.A., Schuck P., Shaginyan V.R. (1997). Fermion condensation and non Fermi liquid behavior in a model with long range forces. Z. Phys. B, 102, 245, DOI: https://doi.org/10.1007/s002570050286
Dykman I.M., Pekar S.I. (1952). Excitons in Ionic Crystals. Dokl. Akad. Nauk. SSSR, 83, 825.
Dykman I.M., Pekar S.I. (1953). Excitons in Ionic Crystals. Trudy Instituta Fiziki AN USSR, 4, 25.
Dykman I.M., Pekar S.I. (1988). Excitons in Ionic Crystals/S.I. Pekar, Selection Works, (Kiev, Naukova dumka, 512 p.)
Eagles D.M. (1969). Possible Pairing without Superconductivity at Low Carrier Concentrations in Bulk and Thin-Film Superconducting Semiconductors. Phys. Rev., 186, 456, DOI: https://doi.org/10.1103/PhysRev.186.456
Ebrahimnejad H., Berciu M. (2012). Trapping of three-dimensional Holstein polarons by various impurities. Phys. Rev. B, 85, 165117, DOI: https://doi.org/10.1103/PhysRevB.85.165117
Edstam J., Olsson H.K. (1994). London penetration depth of YBCO in the frequency range 80–700 GHz. Physica B, 194–196, Part 2, 1589–1590, DOI:https://doi.org/10.1016/0921-4526(94)91294-7
Eliashberg G.M. Interactions between electrons and lattice vibrations in a superconductor. (1960). Sov. Phys. JETP, 11(3),696–702.
Emery V.J., Kivelson S.A. (1995). Importance of phase fluctuations in superconductors with small superfluid density. Nature, 374, 434–437, DOI: https://doi.org/10.1038/374434a0
Emin D. (1986). Self-trapping in quasi-one-dimensional solids. Phys. Rev. B, 33, 3973, DOI: https://doi.org/10.1103/PhysRevB.33.3973
Emin D. (1989). Formation, motion, and high-temperature superconductivity of large bipolarons. Phys. Rev. Lett., 62, 1544, DOI: https://doi.org/10.1103/PhysRevLett.62.1544
Emin D., Ye J., Beckel C.L. (1992). Electron-correlation effects in one-dimensional large-bipolaron formation. Phys. Rev. B, 46, 10710, DOI: https://doi.org/10.1103/physrevb.46.10710
Emin D. (2013). Polarons. (Cambridge: Cambridge Univ. Press)
Emin D. (2017). Dynamic d-symmetry Bose condensate of a planar-large-bipolaron-liquid in cuprate superconductors. Phil. Mag., 97, 2931–2945, DOI: https://doi.org/10.1080/14786435.2017.1354137
Erdmenger J., Kerner P., Müller S. (2012). Towards a holographic realization of Homes' law. J. High Energy Phys., 2012, 21, DOI: https://doi.org/10.1007/JHEP10(2012)021
Feynman R.P. (1955). Slow electrons in a polar crystal. Phys. Rev., 97, 660.
Feynman R. (1972). Statistical Mechanics. (W.A.Benjamin, Inc. Massachusetts, USA)
Firsov Yu.A. (1975). Polarons. (M.: Nauka, Moscow) (in Russian)
Firsov Yu.A., Kabanov V.V., Kudinov E.K., Alexandrov A.S. (1999). Comment on "Dynamical properties of small polarons". Phys. Rev. B, 59, 12132, DOI: https://doi.org/10.1103/PhysRevB.59.12132
Foldy L. (1961). Charged Boson Gas. Phys. Rev. B, 124, 649, DOI: https://doi.org/10.1103/PhysRev.124.649
Franck J.P., Jung J., Mohamed M.A-K., Gygax S., Sproule G.I. (1991). Observation of an oxygen isotope effect in superconducting (Y1–xPrx)$Ba_2Cu_3O_{7-\delta}$. Phys. Rev. B, 44, 5318, DOI: https://doi.org/10.1103/PhysRevB.44.5318

Franck J.P. (1994). 'Experimental studies of the isotope effect in high temperature superconductors', in: Ginsberg D.M. (ed.) Physical Properties of High Temperature Superconductors IV, (World Scientific, Singapore) p.189; DOI: https://doi.org/10.1142/2244

Franz M. (2007). Importance of fluctuations. Nat. Phys., 3, 686–687, DOI: https://doi.org/10.1038/nphys739

Freire H., de Carvalho V.S., Pépin C. (2015). Renormalization group analysis of the pair-density-wave and charge order within the termionic hot-spot model for cuprate supercoducters. Phys. Rev. B, 92, 045132, DOI: https://doi.org/10.1103/PhysRevB.92.045132

Fröhlich H., Pelzer H., Zienau S. (1950). XX. Properties of slow electrons in polar materials. The London, Edinburgh, and Dublin Philosophical Magazine and Journal of Science, 41, 221, DOI: https://doi.org/10.1080/14786445008521794

Fröhlich H. (1954). On the Theory of superconductivity: The one dimensional case. Proceedings of the Royal Society A, 223, (1154): 296–305, DOI: https://doi.org/10.1098/rspa.1954.0116

Furrer A. (2005). 'Neutron Scattering Investigations of Charge Inhomogeneities and the Pseudogap State in High-Temperature Superconductors', in Mingos M.P. (ed) Superconductivity in Complex Systems, Structure and Bonding, Series, Vol. 114, p.361, Müller K.A., Bussmann-Holder A. (eds.) (Springer, Berlin Heidelberg) DOI: 10.1007/b12231

Gaididei Y.B., Loktev V.M. (1988). On a theory of the electronic spectrum and magnetic properties of high-T_c superconductors. Phys. Stat. Sol. (b), 147, 307–319, Doi: https://doi.org/10.1002/pssb.2221470135

Garcia D.R., Lanzara A. (2010). Through a lattice darkly: shedding light on electron – phonon coupling in high T_c cuprates. Adv. Cond. Mat. Phys., 2010, ID 807412, DOI: https://doi.org/10.1155/2010/807412

de Gennes P.-G. (1981). Champ critique d'une boucle supraconductrice ramifiée, C.R. Acad. Sci. Ser. II 292, 279.

Gerlach B., Löwen H. (1988). Absence of phonon-induced localization for the free optical polaron and the corresponding Wannier exciton-phonon system. Phys. Rev. B, 37, 8042, DOI: https://doi.org/10.1103/PhysRevB.37.8042

Gerlach B., Löwen H. (1991). Analytical properties of polaron systems or: Do polaronic phase transitions exist or not? Rev. Mod. Phys., 63, 63, DOI: https://doi.org/10.1103/RevModPhys.63.63

Gerlach B., Luczak F. (1996). Ground-State energy of an exciton-(LO) phonon system in two and three dimensions: general outline and three dimensional case. Phys. Rev. B, 54, 12841–12851, DOI: https://doi.org/10.1103/PhysRevB.54.12841

Gerlach B., Kalina F. (1999). Energy spectrum of the optical polaron at finite total momentum. Phys. Rev. B, 60, 10886–10897, DOI: https://doi.org/10.1103/PhysRevB.60.10886

Ghiringhelli G., Le Tacon M., Minola M., Blanco-Canosa S., Mazzoli C., Brookes N.B., Luca G.M., Frano A., Hawthoru D.G., He F., Loew T., Moretti Sala M., Peeters D.C., Salluzzo M., Schierle E., Sutarto R., Sawatzky C.A., Weschke E., Keimer B., Braicovich L. (2012). Long-range incommensurate charge fluctuations in (Y,Nd) $Ba_2Cu_3O_{6+x}$. Science, 337, 821–825 DOI: 10.1126/science.1223532

Ginzburg V.L., Landau L.D. (1950). On the theory of superconductivity. JETF, 20, 1064–1082, (in Russian)

Ginzburg V.L. (1968). Problem of High Temperature Superconductivity:. Usp.Fiz.Nauk, 95, 91–110, DOI: https://doi.org/10.3367/UFNr.0095.196805g.0091

Ginzburg V.L. (2000). Superconductivity: the day before yesterday — yesterday — today — tomorrow. Phys.Usp., 43, 573–583, DOI: 10.1070/PU2000v043n06ABEH000779

Giorgini S., Pitaevskii L.P., Stringari S. (2008). Theory of ultracold atomic Fermi gases. Rev. Mod. Phys., 80, 1215, DOI: https://doi.org/10.1103/RevModPhys.80.1215

Giubileo F., Roditchev D., Sacks W., Lamy R., Klein J. (2001). Strong coupling and double gap density of states in superconducting MgB_2. Phys. Rev. Lett., 87, 17708, DOI: https://doi.org/10.1209/epl/i2002-00415-5

Giubileo F., Roditchev D., Sacks W., Lamy R., Thanh D.X., Klein J., Miraglia S., Fruchart D., Marcus J., Monod Ph. (2002). Two gap state density in MgB_2: a true bulk property of a proximity effect? Phys. Rev. Lett., 58, 764, DOI: https://doi.org/10.1103/PhysRevLett.87.177008

Glauber R.J. (1963). Photon correlations. Phys. Rev. Lett., 10, 84, DOI: https://doi.org/10.1103/PhysRevLett.10.84

Gor`kov L.P. (1959). Microscopic Derivation of the Ginzburg-Landau Equation in the Theory of Superconductivity. Sov.Phys.JETP, 9, 1364, DOI: http://www.jetp.ac.ru/cgi-bin/e/index/e/9/p1364?a=list

Gor`kov L.P., Kopnin N.B. (1988). High-T_c superconductors from the experimental point of view. Sov. Phys. Usp., 31, 850, DOI:10.1070/PU1988v031n09ABEH005623

Greco A., Zeyner R. (1999). Electronic correlations, electron-phonon interaction, and isotope effect in high-T_c cuprates. Phys. Rev. B, 60, 1296, DOI: https://doi.org/10.1103/PhysRevB.60.1296

Griffin A., Snoke D.W., Stringari S. (eds.) (1996). Bose-Einstein Condensation. (Cambridge U.P., New York).

Grimwall G. (1981). The electron-phonon Interaction in Metals, (North-Holland Publ. Comp., Amsterdam)

Gross E.P. (1955). Small Oscillation Theory of the Interaction of a Particle and Scalar Field. Phys. Rev., 100, 1571, DOI: https://doi.org/10.1103/PhysRev.100.1571

Gross E.P. (1976). Strong coupling polaron theory and translational invariance. Ann. Phys. 99 1 DOI:https://doi.org/10.1016/0003-4916(76)90082-8

Grusdt F., Astrakharchik G.E., Demler E. (2017). Bose polarons in ultracold atoms in one dimension: Beyond the Fröhlich paradigm. New J. Phys., 19, 103035, DOI: https://doi.org/10.1088/1367-2630/aa8a2e

Grüner G. (1994). Density waves in Solids, (Addison-Wesley, Reading)

Grzybowski P., Micnas R. (2007). Superconductivity and Charge-Density Wave Phase in the Holstein Model: a Weak Coupling Limit. Acta Physica Polonica A, 111, 455, DOI: https://doi.org/10.12693/APhysPolA.111.453

Gunnarsson O., Rösch O. (2008). Interplay between electronphonon and Coulomb interactions in cuprates. J. Phys. Condens Matter, 20, 043201, DOI: https://doi.org/10.1088/0953-8984/20/04/043201

Hague J.P., Kornilovitch P.E., Alexandrov A.S. (2008). Trapping of lattice polarons by impurities. Phys. Rev. B, 78, 092302, DOI: https://doi.org/10.1103/PhysRevB.78.092302

Haken H. (1958). Die Theorie des Exzitons im festen Körper. Fortschr. Phys., 6, 271

Haken H. (1973). Quantenfeldtheorie des Festkörpers (B.G.Teubner, Stuttgart)

Hakioglu T., Ivanov V.A., Shumovsky A.S., Tanatar B. (1995). Phonon Squeezing vie correlations in the superconducting electron-phonon interaction. Phys. Rev. B, 51, 15363, DOI: 10.1103/physrevb.51.15363

Hamidian M.N., Edkins S.D., Sang Hyun Joo, Kostin A., Eisaki H., Uchida S., Lawler M.J., Kim E.-A., Mackenzie A.P., Fujita K., Lee Jinho, Seamus Davis J.C. (2016). Detection of a Cooper-pair density wave in $Bi_2Sr_2CaCu_2O_{8+x}$. Nature, 532, 343, DOI: 10.1038/nature17411

Hang Z. (1988a). New type of Cooper pairing in systems with strong electron-phonon interaction. Phys. Rev. B, 37, 7419, DOI: https://doi.org/10.1103/PhysRevB.37.7419

Hang Z. (1988b). Variational ground state of a system with strong electron-phonon interaction. Phys. Rev. B, 38, 11865, DOI: https://doi.org/10.1103/PhysRevB.38.11865

Hang Z. (1988c). Reconsideration of a simple model for bipolarons. Sol. St. Comm., 65, 731–734, DOI:https://doi.org/10.1016/0038-1098(88)90374-2
Hang Z. (1988d). Squeezed polarons in one dimension. Phys. Lett. A, 131, 115–118
Hang Z. (1989). Variational treatment of the strong electron-phonon interaction. J. Phys. Condens. Matter., 1, 1641–1651.
Hai G-Q., Candido L., Brito B., Peeters F. (2018). Electron pairing: from metastable electron pair to bipolaron. J. Phys. Comm., 2, 035017, DOI: https://doi.org/10.1088/2399-6528/aaaee0
Harling O.K. (1970). High-Momentum-Transfer Neutron-Liquid-Helium Scattering Bose Condensation. Phys. Rev. Lett., 24, 1046, DOI: https://doi.org/10.1103/PhysRevLett.24.1046
Hashimoto M., Vishik I.M., He Rui-Hua, Devereaux T.P., Shen Z.-X. (2014). Energy gaps in high-transition-temperature cuprate superconductors. Nat. Phys., 10, 483, DOI: https://doi.org/10.1038/nphys3009
Hattass M., Jahnke T., Schössler S., Czasch A., Schöffler M., Schmidt L.Ph.H., Ulrich B., Jagutzki O., Schumann F.O., Winkler C., Kirschner J., Dörner R., Schmidt-Böcking H. (2008). Dynamics of two-electron photoemission from Cu(111). Phys. Rev. B, 77, 165432, DOI: https://doi.org/10.1103/PhysRevB.77.165432
Heeger A.J., Kivelson S., Schrieffer J., Su W.-P. (1988). Solitons in conducting polymers. Rev. Mod. Phys., 60, 781, DOI: https://doi.org/10.1103/RevModPhys.60.781
Heisenberg W. (1930). Die Selbstenergie des Elektrons. ZS F Phys., 65, 4, DOI: https://doi.org/10.1007/BF01397404
Helm T., Kartsovnik M.V., Bartkowiak M., Bittner N., Lambacher M., Erb A., Wosnitza J., Gross R. (2009). Evolution of the Fermi Surface of the Electron-Dopel High-Temperature Superconductor $Nd_{2-x}Ce_xCuO_4$ Revealed by Shubnikov-de Haas Oscillations. Phys. Rev. Lett., 103, 157002, DOI: 10.1103/PhysRevLett.103.157002.
Higgs P.W. (1964). Broken symmetries and the masses of gauge Bosons. Phys. Rev. Lett., 13, 508–509, DOI: https://doi.org/10.1103/PhysRevLett.13.508.
Höhler G.Z. (1955). Wechselwirkung eines nichtrelativistischen Teilchens mit einem skalaren Feld für mittlere Kopplung. I. Zeitschrift für Physik, 140, 192–214, DOI: https://doi.org/10.1007/BF01349378
Hohenberg P.C., Platzman P.M. (1966). High-Energy Neutron Scattering from Liquid He^4. Phys. Rev., 152, 198, DOI: https://doi.org/10.1103/PhysRev.152.198
Holstein T. (1959a). Studies of polaron motion: part 1. The molecular-crystal model. Ann. Phys., 8, 325, DOI:https://doi.org/10.1016/0003-4916(59)90002-8
Holstein T. (1959b). Studies of polaron motion: Part II. The "small" polaron. Annals of Physics, 8, 343–389, DOI:https://doi.org/10.1016/0003-4916(59)90003-X
Homes C.C., Dordevic S.V., Strongin M., Bonn D.A., Ruixing Liang, Hardy W.N., Seiki Komiya, Yoichi Ando, Yu G., Kaneko N., Zhao X., Greven M., Basov D.N., Timusk T. (2004). A universal scaling relation in high-temperature superconductors. Nature 430, 539 DOI: https://doi.org/10.1038/nature02673
Hoogenboom B.W., Renner Ch., Revaz B., Maggio-Aprile I., Fischer O. (2000). Low-energy structures in vortex core tunneling spectra in$Bi_2Sr_2CaCu_2O_8$. Physica C: Superconductivity, 332, 440, DOI:10.1016/S0921-4534(99)00720-0
Hore S.R., Frankel N.E. (1975). Dielectric response of the charged Bose gas in the random-phase approximation. Phys. Rev. B, 12, 2619, DOI: https://doi.org/10.1103/PhysRevB.12.2619
Hore S.R., Frankel N.E. (1976). Zero-temperature dielectric response of the charged Bose gas in a uniform magnetic field. Phys. Rev. B, 14, 1952, DOI: https://doi.org/10.1103/PhysRevB.14.1952
Hubbard J. (1963). Electron correlations in narrow energy bands. Proc R Soc Lond A, 276, 238, DOI: https://doi.org/10.1098/rspa.1963.0204

Huefner S., Hossain M.A., Damascelli A., Sawatzky G.A. (2008). Two Gaps Make a High Temperature Superconductor? Rep. Progr. Phys., 71, 062501, DOI: 10.1088/0034-4885/71/6/062501
Iadonisi G., Ranninger J., de Filips G. (2006). Polarons in Bulk Materials and Systems with Reduced Dimensionality, (IOS Press, Amsterdam, Oxford, Tokio, Washington D.C.)
Ishiguro T., Yamaji K., Saito G. (1998). Organic Superconductors. (Springer, Berlin)
Iwao S. (1976). Quark-Polaron Model of Fundamental Particles. Lett. Nuovo Cimento, 15, 331.
Iwasawa H., Douglas J.F., Sato K., Masui T., Yoshida Y., Sun Z., Eisaki H., Bando H., Ino A., Arita M., Shimada K., Namatame H., Taniguchi M., Tajima S., Uchida S., Saitoh T., Dessau D.S., Aiura Y. (2008). Isotopic Fingerprint of Electron-Phonon Coupling in High-T_c Cuprates. Phys. Rev. Lett., 101, 157005, DOI: https://doi.org/10.1103/PhysRevLett.101.157005
Izyumov Yu.A. (1997). Strongly correlated electrons: the *t–J* model. Phys.Usp., 40, (5), 445, DOI: 10.1070/PU1997v040n05ABEH000234
Jackson H.W. (1973). Reexamination for a Bose-Einstein condensate in superfluid He^4. Phys. Rev. A, 10, 278–294, DOI: https://doi.org/10.1103/PhysRevA.10.278
Jackson S.A., Platzman P.M. (1981). Polaronic aspects of two-dimensional electrons on films of liquid helium. Phys. Rev. B, 24, 499–502, DOI: https://doi.org/10.1103/PhysRevB.24.499
Junior L.A.R., Stafström S. (2015). Polaron Stability in molecular semiconductors: theoretical insight into the impact of the temperature, electric field and the system dimensionality. Phys. Chem. Chem. Phys., 17, 8973–8982, DOI: https://doi.org/10.1039/C4CP06028H
Kakani S.L., Kakani S. (2009). Superconductivity, (Anshan, Ltd., Kent, UK).
Kamerlingh Onnes H. (1911). The superconductivity of Mercury. Comm. Phys. Lab. Univ. Leiden, 122 and 124.
Kandemir B.S., Altanhan T. (1994). Some properties of large polarons with squeezed states. J. Phys. Condens. Mat., 24, 4505–4514, DOI: https://doi.org/10.1088/0953-8984/6/24/012
Kandemir B.S., Cetin A. (2005). Impurity magnetopolaron in a parabolic quantum dot: the squeezed-state variational approach. J. Phys. Condens. Mat., 17, 667–677, DOI: https://doi.org/10.1088/0953-8984/17/4/009
Kashirina N.I., Lakhno V.D., Sychyov V.V. (2002). Electron Correlations and Spatial Configuration of the Bipolaron. Phys. Status Solidi B, 234, 563–570, DOI:https://doi.org/10.1002/1521-3951(200211)234:2<563::AID-PSSB563>3.0.CO;2-E
Kashirina N.I., Lakhno V.D., Sychyov V.V. (2003) Correlation effects and Pekar bipolaron (arbitrary electron–phonon interaction). Phys. Status Solidi B, 239, 174, DOI: https://doi.org/10.1002/pssb.200301818
Kashirina N.I., Lakhno V.D., Sychyov V.V. (2005). Polaron effects and electron correlations in two-electron systems: Arbitrary value of electron-phonon interaction. Phys. Rev. B, 71, 134301, DOI: https://doi.org/10.1103/PhysRevB.71.134301
Kashirina N.I., Lakhno V.D. (2010). Large-radius bipolaron and the polaron–polaron interaction. Phys. Usp., 53, 431–453, DOI: https://doi.org/10.3367/UFNe.0180.201005a.0449
Kashirina N.I., Lakhno V.D., Tulub A.V. (2012). The virial theorem and the ground state problem in polaron theory. JETP, 114, 867, DOI: https://doi.org/10.1134/S1063776112030065
Kashirina N.I., Lakhno V.D. (2013). Mathematical modeling of autolocalized states in condensed media. (Fizmatlit, Moscow) (in Russian).
Kashirina N.I., Lakhno V.D. (2014). Continuum model of the one-dimensional Holstein bipolaron in DNA. Math. Biol. Bioinform., 9, 430.
Kashirina N.I., Lakhno V.D. (2015). Bipolaron in anisotropic crystals (arbitrary coupling). Math. Biol. &Bioinform., 10, 283, DOI: https://doi.org/10.17537/2015.10.283
Kashurnikov V.A., Krasavin A.V. (2010). Numerical Method in Quantum statistics (M.: Fizmatlit, Moscow) (in Russian).

Kasumov Yu., Kociak M., Guéron S., Reulet B., Volkov V.T., Klinov D.V., Bouchiat H. (2001). Proximity-Induced Superconductivity in DNA. Science, 291, 280 DOI: https://doi.org/10.1126/science.291.5502.280.

Keldysh L.V., Kopaev Yu.V. (1964). Possible instability of semimetallic State Toward Coulomb Interaction. Fiz. Tv. Tela, 6, 279.

Keldysh L.V., Kozlov A.N. (1967). Collective excitations in semiconductors. JETP, 27, N3, 521.

Kervan N., Altanhan T., Chatterjee A.A. (2003). Variational approach with squeezed-states for the polaronic effects in quantum dots. Phys.Lett.A, 315, 280–287, DOI:https://doi.org/10.1016/S0375-9601(03)01011-9

Ketterle W., Zwierlein M.W. (2007). 'Making, probing and understanding ultracold Fermi gases', in: M. Inguscio, W. Ketterle, C. Salomon, eds. Ultra-cold Fermi gases, p.95 IOS Press, Amsterdam.

Khasanov R., Eshchenko D.G., Luetkens H., Morenzoni E., Prokscha T., Suter A., Garifianov N., Mali M., Roos J., Conder K., Keller H. Direct observation of the oxygen isotope effect on the in-plane magnetic field penetration depth in optimally doped $YBa_2Cu_3O_{7-\delta}$. Phys. Rev. Lett., 92, 057602, (2004) DOI: https://doi.org/10.1103/PhysRevLett.92.057602

Kimura H., Noda Y., Goka H., Fujita M., Yamada K., Shirane G. (2005). Soft phonons and structural phase transitions in $La_{1.875}Ba_{0.125}CuO_4$. J. Phys. Soc. Jpn. 74, 445–449, DOI: https://doi.org/10.1143/JPSJ.74.445

Kirtley J.R., Tsuei C.C., Ariando C.J.M. Verwijs, Harkema S., Hilgenkamp H. (2006). Angle-resolved phase sensitive determination of the in-plane gap symmetry in $YBa_2Cu_3O_7$. Nature Physics, 2, 190–194, DOI: https://doi.org/10.10.38/nphys215

Kittel Ch. (1963). Quantum Theory of Solid, (New York-London: J. Wiley)

Kivelson S.A., Rokhsar D.S. (1990). Bogoliubov quasiparticles, spinons, and spin-charged decoupling in superconductors. Phys. Rev. B, 41, 11693, DOI: https://doi.org/10.1103/PhysRevB.41.11693

Klamt A. (1988). Tight-binding polarons. I. A new variational approach to the molecular-crystal model. J. Phys. C: Solid State Phys., 21, 1953, DOI: https://doi.org/10.1088/0022-3719/21/10/014

Klimin S.N., Devreese J.T. (2012). Comments on "Translation-invariant bipolarons and the problem of high-temperature superconductivity". Sol. St. Comm., 152, 1601, DOI: https://doi.org/10.1016/j.ssc.2012.05.013

Klimin S.N., Devreese J.T. (2013). Reply to "On the cutoff parameter in the translation-invariant theory of the strong coupling polaron". Sol. St. Comm., 153, 58, DOI: https://doi.org/10.1016/j.ssc.2012.10.012

Knox S. (1963). Theory of excitons, New York: Academic.

Kogan A., Rak M.S., Vig S., Husain A.A., Flicker F., Joe Y.Il., Venema L., MacDougall G.J., Chiang T.C., Fradkin E., van Wezel J., Abbamonte P. (2017). Signatures of exciton condensation in a transition metal dichalogenide. Science, 358, 1314–1317, DOI: 10.1126/science.aam6432

Kohn W. (1959). Image of the Fermi surface in the vibration spectrum of a metal. Phys. Rev. Lett., 2, 393–394, DOI: 10.1103/PhysRevLett2.393

Korepin V.E., Essler F.H. (eds.) (1994). Exactly solvable models of strongly correlated electrons. Advanced series in math-ematical physics, v 18. World Scientific, Singapore.

Kouzakov K.A., Berakdar J. (2003). Photoinduced emission of cooper pairs from superconductors. Phys. Rev. Lett., 91, 257007, DOI: https://doi.org/10.1103/PhysRevLett.91.257007

Kresin V.Z., Wolf S.A. (2009). Colloquium: electron-lattice interaction and ins impact on high superconductivity. Rev. Mod. Phys., 81, 481–501, DOI: 10.1103/RevModPhys.81.481

Kresin V., Ovchinnikov S., Wolf S. (2021). Superconducting State: Mechanisms and Materials, Oxford Sci. Publ., Oxford, Oxford, UK. 512pp.

Kruchinin S., Nagao H., Aono S. (2011). Modern Aspects of superconductivity. Theory of superconductivity, (World Sci., River Edge, NJ, USA).
Kuper C.G., Whitfield G.D. (1963). Polarons and Excitons (Edinbourgh: Oliver&Boyd Ltd)
Labbé J., Bok J. (1987). Superconductivity in Alcaline-Earth-Substituted La_2CuO_4: A Theoretical Model. Europhys. Lett., 3, 1225, DOI: https://doi.org/10.1209/0295-5075/3/11/012
Lakhno V.D., Nagaev E.L. (1976). Ferron-polaron carrier states in antiferromagnetic semiconductors. Fizika tverdogo tela, 18, 3429–3432.
Lakhno V.D., Nagaev E.L. (1978). Magnetostriction Ferrons. Fizika tverdogo tela, 20, 82–86.
Lakhno V.D., Balabaev N.K. (1983). Selfconsistent solutions in continuum F-center model and the problem of relaxed excited states. Optika i spektroskopiya, 55, 308–312.
Lakhno V.D. (1984). Problem of the ground state of conduction electrons in antiferromagnets subjected to a strong magnetic field. Fizika tverdogo tela, 26, 100–105.
Lakhno V.D. (ed.) (1994). Polarons and Applications (Chichester: Wiley).
Lakhno V.D., Chuev G.N. (1995). Structure of a strongly coupled large polaron. Phys.Usp., 38, 273–285, DOI: 10.1070/PU1995v038n03ABEH000075
Lakhno V.D. (1998). Translation invariance and the problem of the bipolaron. Phys. Usp., 41, 403–406, DOI: https://doi.org/10.1070/PU1998v041n04ABEH000385
Lakhno V.D. (2006). Nonlinear models in DNA conductivity. in Starikov E.B., Lewis J.P., Tanaka S. (eds.) "Modern Methods for Theoretical Physical Chemistry of Biopolymers", Elsevier Science Ltd. Amsterdam, 461–481. DOI: https://doi.org/10.1016/B978-0-444-52220-7.X5062-X.
Lakhno V.D. (2008). DNA nanobioelectronics. Int. J. Quantum Chem., 108, 1970, DOI: https://doi.org/10.1002/qua.21717
Lakhno V.D., Korshunova A.N. (2010). Formation of stationary electronic states in finite homogeneous molecular chains. Math. Biol. Bioinformatics, 5, 1, DOI: 10.17537/2010.5.1 [arXiv: 1305.5732 [cond-mat.other]]
Lakhno V.D. (2010a). Davydov's solitons in a homogeneous nucleotide chain. Int. J. Quant. Chem. 110, 127, DOI: https://doi.org/10.1002/qua.22264.
Lakhno V.D. (2010b). Energy and critical ionic-bond parameter of a 3D-large radius bipolaron. JETP, 110, 811, DOI: https://doi.org/10.1134/S1063776110050122
Lakhno V.D., Sultanov V.B. (2011). On the possibility of bipolaronic states in DNA. Biophysics, 56, 210, DOI: https://doi.org/10.1134/S0006350911020175
Lakhno V. (2011). s-f(d) Exchange Mechanism of Magnon Generation by slow spin polarons. Int. J. Mod. Phys. B, 25, N5, 619–627, DOI: https://doi.org/10.1142/S0217979211057852
Lakhno V.D. (2012a). Translation-invariant bipolarons and the problem of high temperature superconductivity. Solid. State. Commun., 152, 621, DOI: https://doi.org/10.1016/j.ssc.2012.01.013
Lakhno V.D. (2012b). On the cutoff parameter in the translation-invariant theory of the strong coupling polaron (response to comments [8] on the paper V.D. Lakhno, SSC 152 (2012) 621). Sol. St. Comm., 152, 1855, DOI: https://doi.org/10.1016/j.ssc.2012.07.019
Lakhno V.D. (2012c) On the inconsistency of Porsch-Röseler cutoff theory. (reply to [arXiv:1208.1166v2]). *arXiv:1211.0382*
Lakhno V.D. (2013). Translation invariant theory of polaron (bipolaron) and the problem of quantizing near the classical solution. JETP, 116, 892, DOI: https://doi.org/10.1134/S1063776113060083
Lakhno V.D. (2014). Large-radius Holstein polaron and the problem of spontaneous symmetry breaking. Prog. Theor. Exp. Phys., 2014, 073I01, DOI: https://doi.org/10.1093/ptep/ptu075
Lakhno V.D. (2015a). TI-bipolaron theory of superconductivity. arXiv:1510.04527v1 [cond-mat.supr-con]

Lakhno V.D. (2015b). Pekar's ansatz and the strong coupling problem in polaron theory. Phys. Usp., 58, 295, DOI: https://doi.org/10.3367/UFNe.0185.201503d.0317

Lakhno V.D. (2016a). A translation invariant bipolaron in the Holstein model and superconductivity. Springer Plus, 5, 1277, DOI: https://doi.org/10.1186/s40064-016-2975-x

Lakhno V.D. (2016b). Phonon interaction of electrons in the translation-invariant strong-coupling theory. Mod. Phys. Lett. B, 30, 1650031, DOI: https://doi.org/10.1142/S0217984916500317

Lakhno V.D. (2017). Peculiarities in the concentration dependence of the superconducting transition temperature in the bipolaron theory of Cooper pairs. Mod. Phys. Lett. B, 31, 1750125, DOI: https://doi.org/10.1142/S0217984917501251

Lakhno V.D. (2018). Superconducting Properties of 3D Low-density translation-invariant bipolaron gas. Adv. Cond. Matt. Phys., 2018, ID1380986, DOI: https://doi.org/10.1155/2018/1380986

Lakhno V.D. (2019a). Superconducting properties of a nonideal bipolaron gas. Phys. C: Supercond. Its Appl., 561, 1–8, DOI: https://doi.org/10.1016/j.physc.2018.10.009

Lakhno V.D. (2019b). Superconducting properties of 3D low-density TI-bipolaron gas in magnetic field. Condensed Matter, 4, 43, DOI: https://doi.org/10.3390/condmat4020043

Lakhno V.D. (2019c). New method of soft modes investigation by Little-Parks effect. arXiv:1908.05735 [cond-mat.supr-con]

Lakhno V.D. (2020a). Translational-invariant bipolarons and superconductivity. Condensed Matter, 5, 30, DOI: https://doi.org/10.3390/condmat5020030

Lakhno V.D. (2020b). Isotope Effect in the translation-invariant bipolaron theory of high-temperature superconductivity. Condensed Matter., 5, 80, DOI: https://doi.org/10.3390/condmat5040080

Lakhno V.D. (2020c). Translation invariant bipolarons and charge density waves in high-temperature superconductors. Keldysh Inst. Prepr., 57, 1–13, (in Russian) DOI: https://library.keldysh.ru/preprint.asp?id=2020-57, DOI: https://doi.org/10.20948/prepr-2020-57

Lakhno V.D. (2021a). Translation-invariant excitons in a phonon field. Condens. Matter, 6, 20, DOI: https://doi.org/10.3390/condmat6020020

Lakhno V.D. (2021b). Pseudogap isotope effect as a probe of bipolaron mechanism in high temperature superconductors. Materials, 14, 4973, DOI: https://doi.org/10.3390/ma14174973

Lakhno V.D. (2021c). Translation-invariant bipolarons and charge density waves in high-temperature superconductors. Frontiers in Physics, 9, Art. 662926, DOI: 10.3389/fphy.2021.662926

Landau L.D. (1933). On the motion of electrons in a crystal lattice. Phys. Z. Sowjetunion, 3, 644.

Landau L.D. (1941). The theory of superfluidity of helium II. J. Phys. USSR., 5, 71.

Landau L.D. (1947). To the theory of HeII superfluidity. J. Phys. USSR, 11, 91.

Landau L.D., Pekar S.I. (1948). Polaron effective mass. JETF, 18, 419.

Landau L.D., Abrikosov A.A., Halatnikov I.M. (1954). Asimptotic expression for Green function in quantum electrodynamics. DAN SSSR, 95, 1177.

Lanzara A., Zhao G-M., Saini N.L., Bianconi A., Conder K., Keller H., Müller K.A. (1999). Oxygen-isotope shift on the charge-stripe ordering temperature in $La_{2-x}Sr_xCuO_4$ from x-ray absorption spectroscopy. J. of Phys.: Condensed Matter, 11(48), L541–L546, DOI: 10.1088/0953-8984/11/48/103

Larkin A., Varlamov A. (2005). Theory of fluctuations in superconductors (Oxford University Press, Oxford)

Lawler M.J., Fujita K., Lee Jhinhwan, Schmidt A.R., Kohsaka Y., Kim Ch.K., Eisaki H., Uchida S., Davis J.C., Sethna J.P., Kim Eun-Ah. (2010). Intra-unit-cell electronic nematicity of the high-T_c copper-oxide pseudogap states. Nature, 466, 347, DOI: 10.1038/nature09169

Lebed A.G. (ed.) (2008). The physics of organic superconductors and conductors. (Springer series in materials science, Springer, Berlin).

Lee T.D., Low F., Pines D. (1953). The motion of slow electrons in a polar crystal. Phys. Rev., 90, 297, DOI: https://doi.org/10.1103/PhysRev.90.297

Lee P.A., Nagaosa N., Wen X-G. (2006). Doping a Mott insulator: Physics of high-temperature superconductivity. Rev. Mod. Phys., 78, 17, DOI: https://doi.org/10.1103/RevModPhys.78.17

Lee P.A. (2014). Amperean pairing and the pseudogap phase of cuprate superconductors. Phys. Rev. X, 4, 031017, DOI: https://doi.org/10.1103/PhysRevX.4.031017

Levinson I.B., Rashba E.I. (1974). Threshold phenomena and bound states in the polaron problem. Sov.Phys.Usp., 16, 892–912, DOI: 10.1070/PU1974v016n06ABEH004097

Lewis F.D., Wu Y. (2001). Dynamics of superexchange photoinduced electron transfer in duplex DNA. J. Photochem. Photobiol., 2, 1, DOI:https://doi.org/10.1016/S1389-5567(01)00008-9.

Lieb E.H., Yamazaki K. (1958). Ground state energy and effective mass of the polaron. Phys. Rev., 111, 728–733, DOI: https://doi.org/10.1103/PhysRev.111.728

Lieb E.H., Thomas L.E. (1997). Exact ground state energy of the strong coupling polaron. Commun. Math. Phys., 183, 511–519, DOI: https://doi.org/10.1007/s002200050040

Lifshitz E.M., Pitaevskii L.P. (1980) Statistical Physics, Pergamon Press, Oxford, Part 2, v.9

Little W.A., Parks R.D. (1962). Observation of quantum periodicity in the transition temperature of a superconducting cylinder. Phys. Rev. Lett., 9, 9, DOI: https://doi.org/10.1103/PhysRevLett.9.9

Liu Y., Zadorozhny Y., Rosario M.M., Rock B.Y., Carrigan P.T., Wang H. (2001). Destruction of the global phase coherence in ultrathin, doubly connected superconducting cylinders. Science, 294, 2332, DOI: https://doi.org/10.1126/science.1066144

Loktev V.M. (1996). Mechanisms of high-temperature superconductivity of copper oxides. Low Temperature Physics, 22, 3.

London F. (1938). The λ-phenomenon of liquid helium and the Bose-Einstein degeneracy. Nature, 141, 643–644, DOI: https://doi.org/10.1038/141643a0

Low F., Pines D. (1955). Mobility of slow electrons in polar crystals. Phys. Rev., 98, 414, Doi: https://doi.org/10.1103/PhysRev.98.414

Löwen H. (1988). Absence of phase transitions in Holstein systems. Phys. Rev. B, 37, 8661, DOI: https://doi.org/10.1103/PhysRevB.37.8661

Madelung O. (1972). Festkorpertheorie I, II: Springer-Verlag, Berlin, Heidelberg, New York.

Maggio-Aprile I., Renner Ch., Erb A., Walker E., Fischer Ø. (1995). Direct Vortex Lattice Imaging and Tunneling Spectroscopy of Flux Lines on $YBa_2Cu_3O_{7-\delta}$. Phys. Rev. Lett., 75, 2754, DOI: https://doi.org/10.1103/PhysRevLett.75.2754

Mankowsky R., Subedi A., Först M., Mariager S. O., Chollet M., Lemke H. T., Robinson J. S., Glownia J. M., Minitti M. P., Frano A., Fechner M., Spaldin N. A., Loew T., Keimer B., Georges A., Cavalleri A. (2014). Nonlinear lattice dynamics as a basis for enhanced superconductivity in $YBa_2Cu_3O_{6.5}$. Nature 516, 71–73, DOI: https://dx.doi.org/10.1038/nature13875

Manske D. (2004). Theory of Unconventional Superconductors, (Springer, Heidelberg, Germany)

Marchand D.J.J., De Filippis G., Cataudella V., Berciu M., Nagaosa N., Prokof'ev N.V., Mishchenko A.S., Stamp P.C.E. (2010). Sharp transition for single polarons in the one-dimensional Su-Schrieffer-Heeger model. Phys. Rev. Lett., 105, 266605, DOI: https://doi.org/10.1103/PhysRevLett.105.266605

Marouchkine A. (2004). Room-Temperature Superconductivity: Cambridge Int. Sci. Publ., Cambridge.

Marsiglio F., Carbotte J.P. (1991). Gap function and density of states in the strong-coupling limit for an electron-boson system. Phys. Rev. B, 43, 5355, DOI: https://doi.org/10.1103/PhysRevB.43.5355

Matsui H., Sato T., Takahashi T., Wang S.-C., Yang H.-B., Ding H., Fujii T., Watanabe T., Matsud A. (2003). BCS-like Bogoliubov quasiparticles in high-T_c superconductors observed by angle-resolved photoemission spectroscopy. Phys. Rev. Lett., 90, 217002, DOI: https://doi.org/10.1103/PhysRevLett.90.217002

Medicherla V.R.R., Patil S., Singh R.S., Maiti K. (2007). Origin of ground state anomaly in LaB_6 at low temperatures. Appl. Phys. Lett., 90, 062507, DOI: https://doi.org/10.1063/1.2459779

Meevasana W., Devereaux T.P., Nagaosa N., Shen Z-X., Zaanen J. (2006a). Calculation of overdamped c-axis charge dynamics and the coupling to polar phonons in cuprate superconductors. Phys. Rev. B, 74, 174524, DOI: https://doi.org/10.1103/PhysRevB.74.174524

Meevasana W., Ingle N.J.C., Lu D.H., Shi J.R., Baumberger F., Shen K.M., Lee W.S., Cuk T., Eisaki H., Devereaux T.P., Nagaosa N., Zaanen J., Shen Z.-X. (2006b). Doping dependence of the coupling of electrons to bosonic modes in the single-layer high-temperature $Bi_2Sr_2CuO_6$ superconductor. Phys. Rev. Lett., 96, 157003, DOI: https://doi.org/10.1103/PhysRevLett.96.157003

de Mello E.V.L., Ranninger J. (1997). Dynamical properties of small polarons. Phys. Rev. B, 55, 14872, DOI: https://doi.org/10.1103/PhysRevB.55.14872

de Mello E.V.L., Ranninger J. (1999). Reply to "Comment on 'Dynamical properties of small polarons'". Phys. Rev. B, 59, 12135, DOI: https://doi.org/10.1103/PhysRevB.59.12135

Mel'nikov V.I., Volovik G.E. (1974). Polarons in the strong coupling limit. Sov. Phys. JETP, 38 (4),819–823.

Mel'nikov V.I. (1977). Electron-phonon bound states in a one-dimensional system. Sov. Phys. JETP 45(6),1233–1235.

Micnas R., Ranninger J., Robaszkiewicz S. (1990). Superconductivity in narrow-band systems with local nonretarded attractive interactions. Rev. Mod. Phys., 62, 113, DOI: https://doi.org/10.1103/RevModPhys.62.113

Mishchenko A.S. (2005). Diagrammatic Monte Carlo method as applied to the polaron problems. Phys.Usp., 48, 887–902, DOI: 10.1070/PU2005v48n09ABEH002632

Mishchenko A.S., Nagaosa N., Shen Z.-X., De Filippis G., Cataudella V., Devereaux T.P., Bernhard C., Kim K.W., Zaanen J. (2008). Charge dynamics of doped holes in high T_c cuprate superconductors: a clue from optical conductivity. Phys. Rev. Lett., 100, 166401, DOI: https://doi.org/10.1103/PhysRevLett.100.166401

Mishchenko A.S., Nagaosa N., Alvermann A., Fehske H., De Filippis G., Cataudella V., Sushkov O.P. (2009). Localization-delocalization transition of a polaron near an impurity. Phys. Rev. B, 79, 180301, DOI: https://doi.org/10.1103/PhysRevB.79.180301

Misochko O.V. (2003). Electronic Raman scattering in high-temperature superconductors. Phys. Usp., 46, (4), 373, DOI: https://dx.doi.org/10.1070/PU2003V46n04ABEH001257

Misochko O.V. (2013). Nonclassical states of lattice excitations: squeezed and entangled phonons. Phys.Usp., 56, 868–882, DOI: 10.3367/UFNe.0183.201309b.0917

Mitrano M., Cantaluppi A., Nicoletti D., Kaiser S., Perucchi A., Lupi S., Di Pietro P., Pontiroli D., Riccò M., Clark S. R., Jaksch D., Cavalleri A. (2016). Possible light-induced superconductivity in K_3C_{60} at high temperature. Nature 530, 461–464, DOI: https://dx.doi.org/10.1038/nature16522

Miyake S.J. (1975). Strong-coupling limit of the polaron ground state. J. Phys. Soc. Jpn., 38, 181–182, DOI: https://doi.org/10.1143/JPSJ.38.181

Miyake S.J. (1976). The ground state of the optical polaron in the strong-coupling case. J. Phys. Soc. Jpn., 41, 747–752, DOI: https://doi.org/10.1143/JPSJ.41.747

Miyake S.J. (1994). 'Bound Polaron in the Strong-coupling Regime', in Lakhno V.D. (ed.) Polarons and Applications, Wiley, Leeds, p. 219.

Mondal M., Kamlapure A., Chand M., Saraswat G., Kumar S., Jesudasan J., Benfatto L., Tripathi V., Raychaudhuri P. (2011). Phase fluctuations in a strongly disordered **s**-wave NbN superconductor close to the metal-insulator transition. Phys. Rev. Lett., 106, 047001, DOI: https://doi.org/10.1103/PhysRevLett.106.047001

Mook H.A., Yethiraj M., Aeppli G., Mason T.E., Armstrong T. (1993). Polarized neutron determination of the magnetic excitations in $YBa_2Cu_3O_7$. Phys. Rev. Lett., 70(22), 3490, Doi: https://doi.org/10.1103/PhysRevLett.70.3490

Moon E., Sachdev S. (2009). Competition between spin density wave order and superconductivity in the underdoped cuprates. Phys. Rev. B, 80, 035117, DOI: https://doi.org/10.1103/PhysRevB.80.035117

Moriya T., Ueda K. (2000). Spin fluctuations and high temperature superconductivity. Adv. Phys., 49, 555, DOI: https://doi.org/10.1080/000187300412248

Nagaev E.L. (1979). Physics of magnetic semiconductors, (M.: Nauka, Moscow) (in Russan)

Nagy P. (1991). The polaron squeezed states. J. Phys. Condens. Mat., 2, 10573–10579.

Nazarenko F., Dagotto E. (1996). Possible phononic mechanism for $dx_2{-}y_2$ superconductivity in the presence of short-range antiferromagnetic correlations. Phys. Rev., B53, R. 2987, DOI: https://doi.org/10.1103/PhysRevB.53.R2987

Norman M.R., Pines D., Kallin C. (2005). The pseudogap: friend or foe of high T_c? Adv. Phys., 54, 715, DOI: https://doi.org/10.1080/00018730500459906

Novozhilov Yu.V., Tulub A.V. (1957). Functional method in quantum field theory. UFN, 61, 53, DOI: 10.3367/UFNr.0061.195701g.0053

Novozhilov Y.V., Tulub A.V. (1961). The Method of Functionals in the Quantum Theory of Fields (N.Y.: Gordon and Beach).

Nozi`eres P., Schmitt-Rink S. (1985).Bose condensation in an attractive fermion gas: From weak to strong coupling superconductivity. J. Low Temp. Phys., 59, 195, DOI: https://doi.org/10.1007/BF00683774.

Offenhäusser A., Rinaldi R. (eds.) (2009). Nanobioelectronics for electronics, biology, and medicine, (Springer, New York)

Ogg Jr. R.A. (1946). Superconductivity in solid metal-ammonia solutions. Phys. Rev., 70, 93, DOI: https://doi.org/10.1103/PhysRev.70.9

Okahata Y., Kobayashi T., Tanaka K., Shimomura M. (1998). Anisotropic electric conductivity in an aligned DNA cast film. J. Am. Chem. Soc., 120, 6165, DOI: https://doi.org/10.1021/ja980165w

Okomel'kov A.V. (2002). The normal mode spectrum in a two-dimensional lattice of neutral atoms. Physics of Solid State, 44, (10), 1981–1987, DOI: 10.1134/1.1514792

Overend N., Howson M.A., Lawrie I.D. (1994). 3D X-Y scaling of the specific heat of $Y\,Ba_2Cu_3O_7$ single crystals. Phys. Rev. Lett., 72, 3238, DOI: https://doi.org/10.1103/PhysRevLett.72.3238

de Pablo P. J., Moreno-Herrero F., Colchero J., Herrero J. Gómez, Herrero P., Baró A. M., Ordejón Pablo, Soler José M., Artacho, E. (2000). Absence of dc-conductivity in λ-DNA. Phys. Rev. Lett., 85, 4992, DOI: https://doi.org/10.1103/PhysRevLett.85.4992.

Pan S.H., Hudson E.W., Gupta A.K., Ng K.-W., Eisaki H., Uchida S., Davis J. C. (2000). STM studies of the electronic structure of vortex cores in $Bi_2Sr_2CaCu_2O_8$. Phys. Rev. Lett., 85, 1536, DOI: https://doi.org/10.1103/PhysRevLett.85.1536

Panagopoulos C., Cooper J.R., Xiang T. (1998). Systematic behavior of the in-plane penetration depth in d-wave cuprates. Phys. Rev. B, 57, 13422, DOI: https://doi.org/10.1103/PhysRevB.57.13422

Park S.R, Song D.J., Leem C.S., Kim Ch., Kim C., Kim B.J., Eisaki H. (2008). Angle-resolved photoemission spectroscopy of electron-doped cuprate superconductors: isotropic electron-phonon coupling. Phys. Rev. Lett., 101, 117006, DOI: https://doi.org/10.1103/PhysRevLett.101.117006

Pashitskii E.A. (1993). Plasmon mechanism of high-temperature superconductivity in cuprate metal-oxide compounds. JETP, 76, 425.

Pashitskii E.A., Chernousenko V.M. (1971). “Plasmon” Mechanism of Superconductivity in Degenerate Semiconductors and Semimetals. II. JETP, 33, 802.
Pashitskii E.A., Vineckii V.L. (1987). Plasmon and bipolaron mechanisms of high-temperature superconductivity. JETF Lett., 46, 124.
Pashitskii E.A. (2016). The critical temperature as a function of the number of Cooper pairs, and the superconductivity mechanism in a layered LaSrCuO crystal. Low Temp. Phys., 42, 1184, DOI: https://doi.org/10.1063/1.4973010
Patil S., Medicherla V.R.R., Ali K., Singh R.S., Manfrinetti P., Wrubl F., Dhar S.K., Maiti K. (2017). Observation of pseudogap in MgB_2. Journ. of Phys.: Cond. Matter, 29, 465504, DOI: https://doi.org/10.1088/1361-648X/aa8aa2
Pekar S.I. (1946a). Local quantum electron state in ideal ion crystal. JETF, 16, 341.
Pekar S.I. (1946b). Electron autolocalization in dielectric inertial-polarized media. JETF, 16, 335.
Pekar S.I. (1954). Untersuchungen über die Elektronentherie der Kristalle (Berlin, Akad.-Verl.,); Pekar S.I. Research in Electron Theory of Grystals (USA, Department of Commerce, Washington, 25, D.C.: United States Atomic Energy Commision, Division of Technical Information, 1963).
Pekar S.I., Rashba E.I., Sheka V.I. (1979). Free and self-localized Wannier-Mott excitons in ionic crystals and activation energy of their mutual thermal conversion. JETF, 49, 129.
Pépin C., de Carvalho V.S., Kloss T., Montiel X. (2014). Pseudogap, charge order, and pairing density wave at the hot spots in curpate superconducters. Phys. Rev. B, 90, 195207, DOI: https://doi.org/10.1103/PhysRevB.90.195207
Pereg-Barnea T., Turner P.J., Harris R., Mullins G.K., Bobowski J.S., Raudsepp M., Ruixing Liang, Bonn D.A., Hardy W.N. (2004). Absolute values of the London penetration depth in Y $Ba_2Cu_3O_{6+y}$ measured by zero field ESR spectroscopy on Gd doped single crystals. Phys. Rev. B, 69, 184513, DOI: 10.1103/PhysRevB.69.184513
Pieri P., Strinati G.C. (2000). Strong-coupling limit in the evolution from BCS superconductivity to Bose-Einstein condensation. Phys. Rev. B, 61, 15370, DOI: https://doi.org/10.1103/PhysRevB.61.15370.
Pippard A.B. (1950). Field variation of the superconducting penetration depth. Proc. Roy. Soc. (London) A203, 210.
Plakida N.M. (2010). High Temperature Cuprate Superconductors: Experiment, Theory and Applications, (Springer, Heidelberg, Germany)
Plumb N.C., Reber T.J., Koralek J.D., Sun Z., Douglas J.F., Aiura Y., Oka K., Eisaki H., Dessau D.S. (2010). Low-energy (<10mev) feature in the nodal electron self-energy and strong temperature dependence of the Fermi velocity in $Bi_2Sr_2CaCu_2O_{8+\delta}$. Phys. Rev. Lett., 105, 046402, DOI: https://doi.org/10.1103/PhysRevLett.105.046402
Porath D., Bezryadin A., de Vries S., Dekker C. (2000). Direct measurement of electrical transport through DNA molecules. Nature, 403, 635, DOI: https://doi.org/10.1038/35001029
Porsch M., Röseler J. (1967). Recoil effects in the polaron problem. Phys. Status Soliti B, 23, 365, DOI: https://doi.org/10.1002/pssb.19670230138
Proville L., Aubry S. (1998). Mobile bipolarons in the adiabatic Holstein–Hubbard model in one and two dimensions. Phys. D, 113, 307, DOI:https://doi.org/10.1016/S0167-2789(97)00283-2
Pullmann J., Büttner H. (1977). Effective Hamiltonians and bindings energies of Wannier excitons in polar semiconductors. Phys. Rev., 16, 4480, DOI: https://doi.org/10.1103/PhysRevB.16.4480
Radtke R.J., Norman M.R. (1994). Relation of extended Van Hove singularities to high-temperature superconductivity within strong-coupling theory. Phys. Rev., B50, 9554, DOI: https://doi.org/10.1103/PhysRevB.50.9554

Rajaraman R. (1982).Solitons and instantons. An Introduction to Solitons and Instantons in Quantum Field Theory (North-Holland Publ.Company, Amsterdam)
Rameau J.D., Yang H.B., Gu G.D., Johnso P.D. (2009). Coupling of low-energy electrons in the optimally doped $Bi_2Sr_2CaCu_2O_{8+\delta}$ superconductor to an optical phonon mode. Phys. Rev. B, 80, 184513, DOI: https://doi.org/10.1103/PhysRevB.80.184513
Randeria M. (1997). Precursor Pairing Correlations and Pseudogaps. arXiv: cond-mat/9710223 [cond-mat.str-el].
Randeria M., Trivedi N. (1998). Pairing correlations above T_c and pseudogaps in underdoped cuprates. J. Phys. Chem. Sol., 59, 1754, DOI:https://doi.org/10.1016/S0022-3697(98)00099-7
Rashba E.I., Sturge M.D. (eds.) (1982). Excitons, North Holland, Amsterdam, 496 p.
Reznik D. (2010). Giant electron-phonon anomaly in doped La_2CuO_4 and other cuprates. Adv. Cond. Mat. Phys., 2010, ID 523549, DOI: https://doi.org/10.1155/2010/523549
Ribeiro L.A., da Cunha W.F., de Oliveria Neto P.H., Gargano R., Magela e Silva G. (2013). Effects of temperature and electric field induced phase transitions on the dynamics of polarons and bipolarons. New J. Chem., 37, 2829–2836, DOI: https://doi.org/10.1039/C3NJ00602F
Rosenfeld L. (1932). Über eine mögliche Fassung des Diracschen Programms zur Quantenelektrodynamik und deren formalen Zusammenhang mit der Heisenberg-Paulischen Theorie. Zs. f. Phys., 76, 729.
Rossat-Mignod J., Regnault L.P., Vettier C., Bourges P., Burlet P., Bossy J., Henry J.Y., Lapertot G., (1991). Neutron scattering study of the $YBa_2Cu_3O_{6+x}$. Physica C, 185:86, Doi:https://doi.org/10.1016/0921-4534(91)91955-4
Röseler J. (1968). A new variational ansatz in the polaron theory. Phys. Status Solidi B, 25, 311, DOI: https://doi.org/10.1002/pssb.19680250129
Rubio Temprano D., Mesot J., Janssen S., Conder K., Furrer A., Mutka H., Müller K.A. (2000). Large isotope effect on the pseudogap in the high-temperature superconductor $HoBa_2Cu_4O_8$. Phys. Rev. Lett., 84, 1990, DOI: https://doi.org/10.1103/PhysRevLett.84.1990
Sacépé B., Chapelier C., Baturina T.I.,Vinokur V.M., Baklanov M.R., Sanquer M. (2010). Pseudogap in a thin film of a conventional superconductor. Nat. Commun., 1, 140, DOI: https://doi.org/10.1038/ncomms1140
Sacha K., Timmermans E. (2006). Self-localized impurities embedded in a one-dimensional Bose-Einstein condensate and their quantum fluctuations. Phys. Rev. A, 73, 063604, DOI: https://doi.org/10.1103/PhysRevA.73.063604
Sadovskii M.V. (2001). Pseudogap in high-temperature superconductors. Phys. Usp., 44, 515–539, DOI: 10.1070/PU2001v044n05ABEH000902
Sadovskii M.V. (2021). Planckian relaxation delusion in metals. Phys. Usp., 64, 175–190, DOI: https://doi.org/10.3367/UFNe.2020.08.038821
Scalapino D.J. (2012). A common thread: The pairing interaction for unconventional superconductors. Rev. Mod. Phys., 84, 1383, DOI: https://doi.org/10.1103/RevModPhys.84.1383
Seo K., Chen H.-d., Hu J. (2008). Complementary pair-density-wave and d-wave-checkerboard orderings in high-temperature superconductors. Phys. Rev. B, 78, 094510 DOI: https://doi.org/10.1103/PhysRevB.78.094510
Shaginyan V.R., Stephanovich V.A., Msezane A.Z., Japaridze G.S., Popov K.G. (2017). The influence of topological phase transition on the superfluid density of overdoped copper oxides. Phys. Chem. Chem. Phys, 19, 21964, DOI: https://doi.org/10.1039/C7CP02720F
Schafroth M.R. (1955). Superconductivity of a charged ideal Bose gas. Phys. Rev., 100, 463 DOI: https://doi.org/10.1103/PhysRev.100.463
Schleich W.P. (2001). Quantum optics in phase space, (Wiley-VCH, Berlin).

Schmidt V.V. (1997). The Physics of Superconductors, Müller P., Ustinov A.V. (eds.) Springer-Verlag, Berlin-Heidelberg.
Schrieffer J.R., Wen X.-G., Zhang S.-C. (1988). Spin-bag mechanism of high-temperature superconductivity. Phys.Rev.Lett., 60, 944, Doi: https://doi.org/10.1103/PhysRevLett.60.944
Schrieffer J.R., Wen X.-G., Zhang S.-C. (1989). Dynamic spin fluctuation and the bag mechanism of high-T_c. Phys.Rev.B, 39, 11663, Doi: https://doi.org/101103/PhysRevB.39.11663
Schrieffer J.R. (1999). Theory of Superconductivity, (Westview Press, Oxford, UK)
Schrieffer J.R. (ed.) (2007). Handbook of High-Temperature Superconductivity. Theory and Experiment, J.S. Brooks Associated Ed., Springer, 626 p, Springer Science + Business Media, LLC, New York, USA.
Schüttler H-B., Holstein T. (1986). Dynamics and transport of a large acoustic polaron in one dimension. Ann. Phys., 166, 93.
Schüttler H.-B., Pao G.-H. (1995). Isotope effect in d-wave superconductors. Phys. Rev. Lett., 75, 4504, DOI: https://doi.org/10.1103/PhysRevLett.75.4504
Schultz T.D. (1959). Slow electrons in polar crystals: self-energy, mass, and mobility. Phys. Rev., 116, 526, DOI: https://doi.org/10.1103/PhysRev.116.526
Schuster G.B. (2004). Long-Range Charge Transfer in DNA, (Heidelberg: Springer)
Schweber S.S. (1961). An introduction to Relativistic Quantum Field Theory (Harper & Row: New York, NY, USA)
Selyugin O.V., Smondyrev M.A. (1989). Phase transition and Padé approximants for Fröhlich polarons. Phys. Status Solidi B, 155, 155, DOI: https://doi.org/10.1002/pssb.2221550114
Shanenko A.A., Smondyrev M.A., Devreese J.T. (1996). Stabilization of Bipolarons by Polaron Environment. Solid St. Comm., 98, 1091, DOI:https://doi.org/10.1016/0038-1098(96)00077-4
Shaw P.B., Whitfield G. (1978). Vibrational excitations of a one-dimensional electron-phonon system in strong coupling. Phys. Rev. B, 17, 1495, DOI: https://doi.org/10.1103/PhysRevB.17.1495
Shen K.M., Ronning F., Lu D.H., Lee W.S., Ingle N.J.C., Meevasana W., Baumberger F., Damascelli A., Armitage N.P., Miller L.L., Kohsaka Y., Azuma M., Takano M., Takagi H., Shen Z.-X. (2004). Missing quasiparticles and the chemical potential puzzle in the doping evolution of the cuprate superconductors. Phys. Rev. Lett., 93, 267002, DOI: https://doi.org/10.1103/PhysRevLett.93.267002
Shikin V.B., Monarkha Yu.P. (1973). Free electrons on the surface of liquid helium in the presence of external fields. JETP, 38, 373.
Shimamura S., Matsura M. (1983). Internal-motion dependence of self-trapping of a Wannier exciton. Solid State Communications, 45, 547–550, DOI:https://doi.org/10.10161038-1098(83)90133-3
Shumovskii A.S. (1991). Bogolyubov canonical transformation and collective states of bosonic fields. Theor.Math.Phys., 89, 1323–1329, DOI: https://doi.org/10.1007/BF01017828
Silsbee F.B. (1917). Note on electrical conduction in metals at low temperatures. J. Franklin Inst., 184, 111.
Sinha K.P., Kakani S.L. (2002). Fermion local charged boson model and cuprate Superconductors, Proceedings – National Academy of Sciences, India. Section A, Physical Sciences 72, 153.
Smilde H.J.H., Golubov A.A., Rijnders A.G., Dekkers J.M., Harkema S., Blank D.H.A., Rogalla H., Hilgenkamp H. (2005). Admixtures to d-Wave Gap Symmetry in Untwinned $YBa_2Cu_3O_7$ Superconducting Films Measured by Angle-Resolved Electron Tunneling. Phys. Rev. Lett., 95, 257001, DOI: https://doi.org/10.1103/PhysRevLett.95.257001
Smondyrev M.A. (1986). Diagramms in polaron model. Theor.Math.Phys.,68, (1), 653–664, DOI: https://doi.org/10.1007/BF01017794

Smondyrev M.A., Fomin V.M. (1994). 'Pekar-Fröhlich bipolarons'. In: Lakhno V.D. (ed.), Polarons and applications, Proceedings in Nonlinear Science, Wiley, Leeds, UK, 13–70.
Smondyrev M.A., Shanenko A.A., Devreese J.T. (2000). Stability criterion for large bipolarons in a polaron-gas background. Phys. Rev. B, 63, 024302, DOI: https://doi.org/10.1103/PhysRevB.63.024302
Snider E., Dasenbrock-Gammon N., McBride R., Debessai M., Vindana H., Vencatasamy K., Lawler K.V., Salamat A., Dias R.P. (2020). Room-temperature superconductivity in a carbonaceous sulfur hydride. Nature, 586, 373–377, DOI: https://doi.org/10.1038/s41586-020-2801-z
Snoke D., Kavoulakis G.M. (2014). Bose-Einstein condensation of excitons in Cu_2O: progress over 30 years. Reports on Progress in Physics, 77, 116501, DOI: 10.1088/0034-4885/77/11/116501
Sochnikov I., Shaulov A., Yeshurun Y., Logvenov G., Božović I. (2010). Large oscillations of the magne-toresistance in nanopatterned high-temperature super- conducting films. Nature Nanotech. 5, 516, DOI: https://doi.org/10.1038/nnano.2010.111
Somayazulu M., Ahart M., Mishra A.K., Geballe Z.M., Baldini M., Meng Y., Struzhkin V.V., Hemley R.J. (2019). Evidence for Superconductivity above 260 K in Lanthanum Superhydride at megabar pressures. Phys. Rev. Lett., 122(2), 027001, DOI: https://doi.org/10.1103/PhysRevLett.122.027001
Song K.S., Williams R.T. (1996). Self-Trapped Excitons, (second edition), Springer, 401 p., Berlin.
Song Y., Dai P. (2015). 'High-Temperature Superconductors' in Fernandez-Alonzo F., Price D.L. (eds.) Neutron Scattering-Magnetic and Quantum Phenomena, 48, Experimental Methods in the Physical Sciences, Treatise Lucatorto T., Baldwin K., Yates J.T. (eds.) p.145–193, Elsevier, Amsterdam.
Staley N.E., Liu Y. (2012). Manipulating superconducting fluctuations by the Little-Parks-de Gennes effect in ultrasmall Al loops. PNAS, 109(37),14819–14823, DOI: https://doi.org/10.1073/pnas.1200664109
Starikov E.B., Lewis J.P., Tanaka S. (2006). Modern methods for theoretical physical chemistry of biopolimers, (Elsevier: Amsterdam, Boston, Tokyo).
Storm A.J., Van Noort J., de Vries S., Dekker C. (2001). Insulating behavior for DNA molecules between nanoelectrodes at the 100 nm length scale. Appl. Phys. Lett., 79, 3881, DOI: https://doi.org/10.1063/1.1421086
Sumi A. (1977). Phase Diagram of an Exciton in the Phonon Field. J. Phys. Soc. Jpn., 43, 1286–1294, DOI: https://doi.org/10.1143/JTPJ.43.1286
Suprun S.G. Moizhes B.Ya. (1982). Electron correlation effect in Pekar bipolaron formation. Sov. Phys.Solid State, 24, 903.
Svartholm N. (ed.) (1969). Elementary particle physics, Stockholm: Almquist and Wiksell
Tacon M. Lee, Bosak A., Souliou S.M., Dellea G., Loew T., Heid R., Bohnen K-P., Chiringhelli G., Krisch M., Keimer B. (2014). Inelastic X-ray scattering in $YBa_2Cu_3O_{6,6}$ reveals giant phonon anomalies and elastic central peak due to charge-density wave formation. Nat. Phys., 10, 52–58, DOI: https:// doi.org/10.1038/nphys2805
Taillefer L. (2006). High-temperature superconductivity: Electrons scatter as they pair. Nature Physics, 2, 809–810, DOI: https://doi.org/10.1038/nphys478a
Takada Y. (1978). Plasmon mechanism of superconductivity in two and three-dimensional electron systems. J. Phys. Soc. Japan, 15,786–794, Doi: https://doi.org/10.1143/JPSJ.45.786
Tallon J.L., Islam R.S., Storey J., Williams G.V.M., Cooper J.R. (2005). Isotope effect in the superfluid density of high-temperature superconducting cuprates: stripes, pseudogap, and impurities. Phys. Rev. Lett., 94, 237002, DOI: https://doi.org/10.1103/PhysRevLett.94.237002

Tempere J., Casteels W., Oberthaler M.K., Knoop S., Timmermans E., Devreese J.T. (2009). Feynman path-integral treatment of the bec-impurity polaron. Phys. Rev. B, 80, 184504, DOI: https://doi.org/10.1103/PhysRevB.80.184504

Teich M.C., Saleh B.E.A. (1991). Squeezed states of light. Usp.Fiz.Nauk, 161, (4), 101–136, DOI: 10.3367/UFNr.0161.199104d.0101

Thakur S., Biswas D., Sahadev N., Biswas P.K., Balakrishnan G., Maiti K. (2013). Complex spectral evolution in a BCS superconductor, ZrB_{12}. Sci. Rep., 3, 3342, DOI: https://doi.org/10.1038/srep03342

Timusk T., Statt B. (1999). The pseudogap in high-temperature superconductors: an experimental survey. Rep. Progr. Phys., 62, 61–122, DOI: https://doi.org/10.1088/0034-4885/62/1/002

Tinkham M. (1975). Introduction to Superconductivity, McGraw-Hill Book Company, New York.

Tisza L. (1938). Transport Phenomena in Helium II. Nature, 141, 913, DOI: https://doi.org/10.1038/141913a0

Tkach M.V., Seti J.O., Voitsekhivska O.M., Pytiuk O.Y. (2015). Renormalized energy of ground and first excited state of Fröhlich polaron in the range of weak coupling. Condensed Matter Physics, 18, 33707,1–12.

Tohyama T. (2012). Recent progress in physics of high-temperature superconductors. Jpn. J. Appl. Phys., 51, 010004, DOI: https://doi.org/10.1143/JJAP.51.010004

Toyota N., Land M., Müller J. (2007). Low dimensional molecular metals. Springer series in solid-state sciences, 154. Springer; GmbH & Co., Berlin.

Toyozawa Y. (1961). Self-trapping of an electron by the acoustical mode of lattice vibration. I. Progr. Theor. Phys., 26, 29, DOI: https://doi.org/10.1143/PTP.26.29

Tulub A.V. (1958). Phonon interactions of electrons in polar crystals, JETP, vol. 7, N6, p. 1127.

Tulub A.V. (1960). Accounting the recoil in nonrelativistic quantum field theory. Vestnik Lening. Univ.,Sect.4 Phys.Chem., 15, 104–118, (in Russian).

Tulub A.V. (1961). Slow electrons in polar crystals ZhETF, 41,1828; Sov Phys JETP, 14, 1301, (1962).

Tulub A.V. (2015). Comments on polaron-phonon scattering theory. Theor.Math.Phys., 185, 1533–1546, DOI: https://doi.org/10.1007/s11232-015-0363-2

Tyablikov S.V. (1951). Adiabatic perturbation theory in the problem of particle interaction with quantum field. J.Exp.Theor.Phys., 21, 377.

Tyablikov S.V. (1952). Electron energy spectrum in polaron crystal. J.Exp.Theor.Phys., 23, 381.

Tyablikov S.V. (1967). Methods in the Quantum Theory of Magnetism (Springer: Berlin, Germany, 354 p.).

Uemura Y.J., Luke G.M., Sternlieb B.J., Brewer J.H., Carolan J.F., Hardy W.N., Kadono R., Kempton J.R., Kiefl R.F., Kreitzman S.R., Mulhern P., Riseman T.M., Williams D.Ll., Yang B.X., Uchida S., Takagi H., Gopalakrishnan J., Sleight A.W., Subramanian M.A., Chien C.L., Cieplak M.Z., Xiao Gang, Lee V.Y., Statt B.W., Stronach C.E., Kossler W.J., Yu X.H. (1989). Universal correlations between T_c and n_s/m^* (carrier density over effective mass) in high-T_c cuprate superconductors. Phys. Rev. Lett., 62, 2317, DOI: https://doi.org/10.1103/PhysRevLett.62.2317

Uemura Y.J., Le L.P., Luke G.M., Sternlieb B.J., Wu W.D., Brewer J.H., Riseman T.M., Seaman C.L., Maple M.B., Ishikawa M., Hinks D.G., Jorgensen J.D., Saito G., Yamochi H. (1991). Basic similarities among cuprate, bismuthate, organic, Chevrel-phase, and heavy-fermion superconductors shown by penetration-depth measurements. Phys. Rev. Lett., 66, 2665, DOI: https://doi.org/10.1103/PhysRevLett.66.2665

Uemura Y.J. (1997). Bose-Einstein to BCS crossover picture for high-Tc cuprates. Phys. C Supercond., 282–287,194, DOI: https://doi.org/10.1007/3-540-46511-1_12

Val'kov V.V., Dzebisashvili D.M., Korovushkin M.M., Barabanov A.F. (2021). Spin-polaron concept in the theory of normal and superconducting states cuprates. Physics-Uspekhi, 64,641–670, Doi: https://doi.org/10.3367/UFNe.2020.08.038829

Vansant P., Smondyrev M.A., Peeters F.M., Devreese J.T. (1994). Exact equations for bipolarons in the strong-coupling limit. J. Phys. A, 27, 7925, DOI: https://doi.org/10.1088/0305-4470/27/23/035

Varelogiannis G. (1998). Orthorhombicity mixing of s-and d-gap components in $YBa_2Cu_3O_7$ without involving the chains. Phys. Rev. B, 57, R732, DOI: https://doi.org/10.1103/PhysRevB.57.R732

Varma C.M., Schmitt-Rink S., Abrahams E. (1987). Charge transfer excitations and superconductivity in "ionic" metals. Sol. St. Comm., 62, 681, Doi:https://doi.org/10.1016/0038-1098(87)90407-8

Verbist G., Peeters F.M., Devreese J.T. (1991). Large bipolarons in two and three dimensions. Phys. Rev. B, 43, 2712, DOI: https://doi.org/10.1103/PhysRevB.43.2712

Verbist G., Smondyrev M.A., Peeters F.M., Devreese J.T. (1992). Strong-coupling analysis of large bipolarons in two and three dimensions. Phys. Rev. B, 45, 5662, DOI: https://doi.org/10.1103/PhysRevB.45.5262

Veta M., Kanzaki H., Toyozawa K., Hanamura E. (1986). Excitonic Processes in Solida, Springer, Berlin.

Vidmar L., Bonča J., Maekawa S., Tohyama T. (2009). Bipolaron in the t–J model coupled to longitudinal and transverse quantum lattice vibrations. Phys. Rev. Lett., 103, 186401, DOI: https://doi.org/10.1103/PhysRevLett.103.186401

Vignolle B., Carrington A., Cooper R.A., French M.M.J. (2008). Quantum oscillations in an overdoped high-Tc superconductor. Nature, 455, 952, DOI: 10.1038/nature07323

Vinetsky V.L., Pashitsky E.A. (1975). Superfluidity of charged Bose gas and Bipolaron mechanism of superconductivity. Ukr.Fiz.Zh., 20, 338.

Vinetsky V.L, Meredov O., Yanchuk V.A. (1989). Quantum chemistry of bipolaron in isotrope continuum media. Theor. Exp. Chem., 25, 591.

Vinetsky V.L., Kashirina N.I., Pashitsky E.A. (1992). Bipolaron states of great radius and the problem of high-temperature superconductivity. Ukr.Fiz.Zh., 37, 76–94.

Vishik I.M., Lee W. S., He R-H, Hashimoto M., Hussain Z., Devereaux T.P., Shen Z-X. (2010). ARPES studies of cuprate Fermiology: superconductivity, pseudogap and quasiparticle dynamics. New J. Phys., 12, 105008, DOI: 10.1088/1367-2630/12/10/105008

Vishik I.M., Lee W.S., Schmitt F., Moritz B., Sasagawa T., Uchida S., Fujita K., Ishida S., Zhang C., Devereaux T.P., Shen Z.X. (2010). Doping-dependendent nodal Fermi velocity of the high-temperature superconductor $Bi_2Sr_2CaCu_2O_{8+\delta}$ reveald using high-resolution angle-resolved photoemission spectroscopy. Phys. Rev. Lett., 104, 207002, DOI: https://doi.org/10.1103/PhysRevLett.104.207002

Voityuk A.A. (2005a). Charge transfer in DNA: Hole charge is confined to a single base pair due to solvation effects. J. Chem. Phys., 122, 204904, DOI: https://doi.org/10.1063/1.1924551

Voityuk A.A. (2005b). Are radical cation states delocalized over GG and GGG hole traps in DNA? J. Phys. Chem., 109, 10793, DOI: https://doi.org/10.1021/jp050985c

Wakimoto S., Lee S., Gehring P.M., Birgeneau R.J., Shirane G. (2004). Neutron scattering study of soft phonons and diffuse scattering in insulating $La_{1.95}Sr_{0.05}CuO_4$. J. Phys. Soc. Jpn., 73, 3413–3417, DOI: https://doi.org/10.1143/JPSJ.73.3413.

Wang Y., Agterberg D.F., Chubucov A. (2015a). Interplay between pair-and charge-density-wave orders in underdoped cuprates. Phys. Rev. B, 91, 115103, DOI: https://doi.org/10.1103/PhysRevB.91.115103

Wang Y., Agterberg D.F., Chubucov A. (2015b). Coexistence of charge-density-wave and pair-density -wave orders in underdoped cuprates. Phys. Rev. Lett., 114, 197001, DOI: https://doi.org/10.1103/PhysRevLett.114.197001

Weber W. (1988). A Cud-d excitation model for the paring in the high-T_c cuprates. Z. Physik B - Condensed Matter, 70, 323, Doi: https://doi.org/10.1007/BF01317238

Weisskopf V.F. (1981). The formation of cooper pairs and the nature of superconducting currents. Contemporary Physics., 22, 375–395, DOI: 10.1080/001075181082315

Wentzel G. (1942). Zur Paartheory der Kernkrafter. Helv. Phys. Acta, 15, 111.

Williams J.M., Ferraro J.R., Thorn R.J., Carlson K.D., Geiser U., Wang H.H., Kini A.M., Whangbo M.H. (1992). Organic Superconductors (Including Fullerenes: Synthesis, Structure, Properties, and Theory), (Prentice Hall, Englewood Cliffs).

Wu Dong-Ho, Sridhar S. (1990). Pinning forces and lower critical fields in Y $Ba_2Cu_3O_y$ crystals: Temperature dependence and anisotropy. Phys. Rev. Lett., 65, 2074, DOI: https://doi.org/10.1103/PhysRevLett.65.2074

Wu T., Mayaffre H., Krämer S., Horvatič M., Berthier C., Hardy W.N., Liang R., Bonn D.A., Julien M.-H. (2011). Magnetic-field-induced charge-stripe order in the high temperature superconductor $YBa_2Cu_3O_4$. Nature, 477, 191–194, DOI: https://doi.org/10.1038/nature10345

Wu T., Mayaffre H., Krämer S., Horvatič M., Berthier C., Kuhns P.L., Reyes A.P., Liang R., Hardy W.N., Bonn D.A., Julien M-H. (2013). Emegence of charge order from the vortex state of high temperature superconductor. Nat. Commun., 4, 2113, DOI: 10.1038/ncomms 3113

Wu T., Mayaffre H., Krämer S., Horvatič M., Berthier C., Hardy W.N., Liang R., Bonn D.A., Julien M-H. (2015). Incipient charge order observed by NMR in the normal state of $YBa_2Cu_3O_y$. Nat. Commun., 6, 6438 DOI: 10.1038/ncomms7438

Yelland E.A., Singleton J., Mielke C.H., Harrison N., Balakirev F.F., Dabrowski B., Cooper J.R. (2008). Quantum oscillations in the underdoped cuprate $YBa_2Cu_4O_8$. Phys. Rev. Lett., 100, 047003, DOI: 10.1103/PhysRevLett.100.047003

Yokoya T., Chainani A., Kiss T., Shin S., Hirata K., Kameda N., Tamegai T., Nishio T., Uwe H. (2002). High-resolution photoemission study of low-T_c superconductors: Phonon-induced electronic structures in low-T_c superconductors and comparison with the results of high-T_c cuprates. Physica C, 378–381, 97–101, DOI:https://doi.org/10.1016/S0921-4534(02)01389-8

Yoo K.-H., Ha D. H., Lee J.-O., Park J. W., Kim Jinhee, Kim J. J., Lee H.-Y., Kawai T., Han Yong Choi. (2001). Electrical conduction through poly(dA)-Poly(dT) and poly(dG)-poly(dC) DNA molecules. Phys. Rev. Lett., 87, 198102, DOI: https://doi.org/10.1103/PhysRevLett.87.198102

Zaanen J. (2004). Superconductivity: why the temperature is high. Nature 430, 512, DOI: https://doi.org/10.1038/430512a

Zaanen J. (2016). Condensed-matter physics: Superconducting electrons go Missing. Nature, 536, 282, DOI: https://doi.org/10.1038/536282a

Zech D., Keller H., Conder K., Kaldis E., Liarokapis E., Poulakis N., Müller K.A. (1994). Site-selective oxygen isotope effect in optimally doped $YBa_2Cu_3O_{6+x}$. Nature, 371, 681–683, DOI: https://doi.org/10.1038/371681a0

Zelli M., Kallin C., Berlinksy A.J. (2012). Quantium oscillations in a π-striped superconductor. Phys. Rev. B, 86, 104507, DOI: https://doi.org/10.1103/PhysRevB.86.104507

Zhang Yan-Min, Ze Cheng. (2007). Study of two-mode squeezed magnetopolarons. Commun. Theor. Phys., 47, 747–751.

Zhao G.-M., Hunt M., Conder K., Keller H., Müller K. (1997). Oxygen isotope effects in the manganites and cuprates: Evidence for polaronic charge carriers. Phys. C Supercond., 282–287 (Pt 1), 202–205.

Zhou X.J., Yoshida T., Lanzara A., Bogdanov P.V., Kellar S.A., Shen K.M., Yang W.L., F. Ronning F., Sasagawa T., Kakeshita T., Noda T., Eisaki H., Uchida S., Lin C.T., Zhou F., Xiong J.W., Ti W.X.,

Zhao Z.X., Fujimori A., Hussain Z., Shen Z.-X. (2003). Universal nodal Fermi velocity. Nature, 423, 398, DOI: https://doi.org/10.1038/423398a
Zhou P., Chen L., Liu Y., Sochnikov I., Bollinger A.T., Han M.-G., Zhu Y., He X., Božović I., Natelson D. (2019). Electron pairing in the pseudogap state revealed by shot noise in copper oxide junctions. Nature, 572, 493–496, DOI: https://doi.org/10.1038/s41586-019-1486-7
Ziman J.M. (1960). Electrons and phonons. (Clarendon Press, Oxford).
Zwerger W (ed.) (2012). The BCS-BEC Crossover and the Unitary Fermi Gas Lecture Notes in Physics: Springer, (Berlin, Heidelberg).

Index

https://doi.org/10.1515/9783110786668-023

www.ingramcontent.com/pod-product-compliance
Lightning Source LLC
Chambersburg PA
CBHW061406210326
41598CB00035B/6119